T0338653

"*Brain Fever* by Richard Moxon is a fascinating read of the multi-century history of meningitis and the nearly 400-year effort by multiple generations of scientists to develop effective treatments. The collaborations between Richard, Rino Rappuoli, Ham Smith and myself that resulted in the Meningitis B vaccine (Bexsero) are some of the most rewarding of my scientific life. Just in the short time since its release in 2014, tens of thousands of lives have been saved. *Brain Fever* is the exciting story of how effective vaccines came to be and why they are so critical to humanity."

Craig Venter
American Biotechnologist, Founder of The Institute for Genome Research, Chief Executive Officer of the J. Craig Venter Institute

"This is a wonderful book that recounts the story of one of the great figures in vaccinology, Richard Moxon. A pioneer in the field whose work led directly to several of the most important vaccines for meningitis, Moxon tells the story of how this field developed over his career, utilising a range of tools such as genomics to better discover powerful immunogens. His story bridges continents and many areas of science, from basic to translational. His contributions to the field are reflected in the book, as is his role in the program that produced one of the major COVID-19 vaccines. It is an engaging story about a leading scientist and his contribution to this most important field."

John Bell
Regius Professor of Medicine, University of Oxford

"*Brain Fever* is a timely, compelling and vivid narrative of scientific endeavour that perfectly sets the stage for the ambitious World Health Organisation (2020) roadmap to defeat meningitis by 2030. Richard Moxon's observations, by someone who was deeply involved in many of the key pieces of the puzzle, reveal his fascination with the remarkable people who worked tirelessly to understand and control meningitis and his own fight against the intriguing bacteria that cause the disease."

Andrew Pollard
Professor of Paediatric Infection and Immunity, Director of the Oxford Vaccine Group, University of Oxford

BRAIN FEVER

How Vaccines Prevent Meningitis and Other Killer Diseases

BRAIN FEVER

How Vaccines Prevent Meningitis and Other Killer Diseases

Richard Moxon
University of Oxford, UK

World Scientific

NEW JERSEY · LONDON · SINGAPORE · BEIJING · SHANGHAI · HONG KONG · TAIPEI · CHENNAI · TOKYO

Published by

World Scientific Publishing Europe Ltd.
57 Shelton Street, Covent Garden, London WC2H 9HE
Head office: 5 Toh Tuck Link, Singapore 596224
USA office: 27 Warren Street, Suite 401-402, Hackensack, NJ 07601

Library of Congress Cataloging-in-Publication Data
Names: Moxon, E. Richard, author.
Title: Brain fever : how vaccines prevent meningitis and other killer diseases /
    E. Richard Moxon, University of Oxford, UK.
Description: New Jersey : World Scientific, [2021]
Identifiers: LCCN 2021005217 | ISBN 9781786349873 (hardcover) |
    ISBN 9781800610019 (paperback) | ISBN 9781786349880 (ebook) |
    ISBN 9781786349897 (ebook other)
Subjects: LCSH: Meningitis--Treatment. | Vaccination.
Classification: LCC RC376 .M69 2021 | DDC 616.8/2--dc23
LC record available at https://lccn.loc.gov/2021005217

British Library Cataloguing-in-Publication Data
A catalogue record for this book is available from the British Library.

For any available supplementary material, please visit
https://www.worldscientific.com/worldscibooks/10.1142/Q0291#t=suppl

Desk Editors: Balamurugan Rajendran/Michael Beale/Shi Ying Koe

Typeset by Stallion Press
Email: enquiries@stallionpress.com

Printed in Singapore

To my "Peruvian Princess," our children and grandchildren.

Foreword

I am deeply honoured to write the forward to this wonderful book *Brain Fever*, which charts the career and contributions of one of the greatest clinical scientists of his generation. Richard Moxon is someone who changed science and the world with a remarkable vision, a brilliant mind, steely determination and by working with and inspiring a generation of others to believe they could go further than they themselves thought possible and hence make a difference to millions of people's lives.

The book combines the drama of a detective novel, the cliff edge of a thriller, deep disappointments matched with moments of sheer joy, and the complexities of moving from discovery science, through clinical medicine, to the politics of acceptance and global rollout of vaccines to all those who could benefit from them. *Brain Fever* is an extraordinary story of how the impossible can become reality, and its release in 2021 could not be more timely or more prescient.

The gripping story is told, with self-effacing humour and candour by someone who epitomises the rich historical tradition of clinical scientists. An Edward Jenner for the twentieth and twenty-first centuries. Richard has been able to combine a deep understanding and appreciation of fundamental discovery science, the vision to translate that science into innovative medicines that change people's lives and the hard-nosed diplomatic and political skills to make damn sure vaccines for meningitis would be available to people all around the world. A polymath, in an era of over-specialisation.

SARS-CoV-2 has dominated the world since it emerged in late 2019 and affected every continent and every aspect of life. A reminder, if it were needed, of the eternal challenge of infectious diseases, the importance of vaccines and the vulnerability and interconnectivity of our world. SARS-CoV-2 is also a reminder of the unselfish legacy that Richard inspired. The Jenner Institute in Oxford, which Richard helped establish, is now home to a remarkable group of clinical scientists and an amazing portfolio of vaccines against many of the world's most important and seemingly intractable infectious diseases. In addition to Richard's own seminal work on *Haemophilus influenzae* and *Neisseria meningitidis,* and because of the mentorship and inspiration he provided to others, there are now vaccines for *Salmonella typhi,* Ebola, Middle East Respiratory Syndrome (MERS) and SARS-CoV-2 as well as vaccines in development for tuberculosis and malaria and many more. An incredible legacy.

There is no doubt that with rapidly changing ecology, urbanisation, climate change, increased travel, and fragile public health systems that infectious diseases will always be with us and epidemics will become more frequent, more complex, and harder to prevent and contain. These epidemics will be caused by pathogens we know about, and some we do not yet know of, that will emerge from animals, plants, or the environment. Our changing climate will change the epidemiology of pathogens, their vectors, and the infections they cause; hence the critical importance of sustained commitment to basic science and the ability to translate that science into vaccines that can prevent illness and save lives.

This is an inspiring story of hope and of dreams. Dreams that individuals and teams can make the impossible possible and that science can provide us with solutions to apparently intractable problems. *Brain Fever* and Richard Moxon show us that we can also ensure that science is translated into vaccines that are equitably available and accessible to people all around the world.

Sir Jeremy Farrar
Director of Wellcome

Acknowledgements

There are so many people that I wish to thank. Martin Rowe of Lantern Publications, New York, encouraged me to write this book and provided much helpful advice in preparing an outline. I am deeply grateful to the late, incomparably talented literary agent Felicity Bryan, whose encouragement and advice were hugely important. This included putting me in touch with Candida Brazil, whose experience and literary skills were invaluable in critiquing early drafts. Stephen Soehnlen of World Scientific Publishing Company (WSPC) was from the outset enthusiastic about my book and has been an enduring, ever-patient source of support and advice. I owe a huge debt to Michael Beale and the WSPC team, including Emese Csikai and Suyoung Lee, for their energy and expertise in critiquing proofs, printing and promoting the book. I am especially indebted to Giorgio Corsi whose expertise, ideas and patience have been exemplary in preparing all the diagrams and figures in this book.

Many people have read Chapter drafts and contributed thoughtful and important suggestions. James Morris and Christopher Winearls undertook to read the book in its entirety when it was close to completion and provided many thoughtful and important comments. I received valuable suggestions from Flora Bagenal, Steve Black, Jon Burrough, Sarah Caro, Andrew Cox, Gordon Dougan, Tom Downes, Jon Drori, Andrew Gorringe, Carol Graham, Dan Granoff, the late Tessa John, August and Elizabeth Maffry, Martin Maiden, Duccio Medini, Christopher and Sarah Moxon, Staffan Normark, Paul Nurse, Ian Phillips, Mariagrazia Pizza, Emma Plested, Stan Plotkin, Rino Rappuoli, Jim Richards, Hannah Robinson, Helen Rowe, Philippe Sansonetti, Hamilton Smith, Vinny Smith, Matthew Snape, Chris Tang, Fred Taylor, Jeffrey Weiser, the late Sir David Weatherall and Samantha Vanderslott. I am also hugely indebted to Sir Jeremy Farrar who graciously agreed to write the Foreword.

I am hugely indebted to my family for their input and encouragement. Timothy Moxon read and critiqued early and later versions, set up my blog, *MoxForum*, and has helped his octogenarian father appreciate the enormous importance and reach of social media. No words can do justice or adequately express the gratitude I owe to my wife, Marianne, for the hours she has invested in advising and supporting me on all aspects of the book; her contribution has been monumental.

Throughout my career as a clinician-scientist, I have been fortunate to have the mentorship and friendship of numerous inspiring medical and scientific colleagues at Cambridge University, St. Thomas's Hospital, Boston Children's Hospital, Johns Hopkins University Medical School and the Medical Sciences Division of Oxford University, especially my colleagues in the Department of Paediatrics, the Weatherall Institute of Molecular Medicine (special thanks to the late Peter Butler Derek Hood, Duncan Maskell, Mike Jennings, Joyce Plested and Mary Deadman), the Oxford Vaccine Group and my colleagues in the Jenner Institute. I have had the privilege of teaching and learning from an extraordinary succession of students and trainees. Thanks also to many scientific colleagues in Siena, Italy, with whom I have had such a valuable and stimulating association over many years. Finally, I wish to thank my colleagues at Jesus College, Oxford, for their scholarship and friendship.

Contents

Timeline

1650	First observation of bacteria by Anton van Leeuwenhoek.
1768	Robert Whytt (Edinburgh) describes dropsy of brain, later recognised as meningitis caused by tuberculosis (TB).
1798	Jenner's research on cowpox that pioneered vaccination.
1805	Outbreak of meningitis in Eaux-Vives (Geneva), Switzerland, described by Gaspard Vieusseux.
1845	Theodore Swann uses heat sterilisation to prevent putrefaction.
1850	Florence Nightingale champions hygiene and ventilation in hospitals.
1855	John Snow halts a cholera epidemic by disconnecting the Broad Street water pump in Soho.
1857	Pasteur demonstrates that fermentation is caused by living organisms.
1861	Pasteur and Koch propose germ theory of disease.
1867	Joseph Lister pioneers use of phenol to disinfect surgical wounds.
1879	First live attenuated bacterial vaccine against chicken cholera (Pasteur).
1881	Discovery of pneumococcus (Sternberg, Pasteur). Pasteur demonstrates protection against anthrax in sheep by immunisation.
1884	Koch and Loeffler propose criteria for causal relationship between microbe and disease.
1887	Isolation of meningococcus (*Neisseria meningitidis*) from cerebrospinal fluid (Weichselbaum).
1891	First use of lumbar puncture by Heinrich Quincke.

1893	Pfeiffer proposes *Bacillus influenzae* to be the cause of influenza.
1907	Flexner reduces mortality meningococcal meningitis using serum treatment.
1911	Martha Wollstein performs classic studies on *Bacillus influenzae*.
1914	Results of first effective pneumococcal vaccine (South Africa Gold miners) published.
1918–1919	Pandemic influenza (Spanish flu) widely, but mistakenly, considered to be caused by *B. influenzae*.
1922	Tom Rivers describes 22 cases of *B. influenzae* meningitis.
1923	*B. influenzae* renamed *Haemophilus influenzae*.
1924	Avery and Heidelberger identify capsular polysaccharides of pneumococcus and their role in virulence.
1928	Fred Griffith conducts experiments on pneumococci showing transformation of capsular polysaccharides.
1930	Margaret Pittman describes different capsular types of *H. influenzae*.
1933	Andrewes, Smith and Laidlaw demonstrate that influenza is caused by a virus, not *H. influenzae*. Leroy Fothergill and Joyce Wright show that immunity to *H. influenzae* type b meningitis is mediated by serum factors (later shown to be antibodies).
1934	First reports of capsular polysaccharide of meningococcus (Geoffrey Rake of Rockefeller Institute).
1937	Sulphonamides found to be effective in treating bacterial meningitis.
1939	Hattie Alexander shows that passive immunisation with immune sera reduces mortality of *H. influenzae* meningitis.
1944	Avery, MacLeod and McCarty propose that nucleic acids are the basis of heredity. First use of penicillin as effective treatment against meningitis.
1945	Publication of effectiveness of pneumococcal capsular polysaccharide vaccine in armed forces developed by MacLeod and Heidelberger.
1953	Grace Leidy demonstrates DNA transformation of *H. influenzae*.
1963	Watson and Crick publish the double helical structure of DNA.

1968	First use of polysaccharides to prevent meningococcal meningitis (serogroups A and C) in armed forces by Gotschlich and Artenstein.
1972	Isolation of *H. influenzae* type b capsular polysaccharide and development of assays to detect type-specific antibodies. Meningococcus B polysaccharide shown to be a very weak immunogen in humans.
1973	Isolation of antibiotic-resistant *H. influenzae* type b from cases of meningitis.
1974	Development of animal models of *H. influenzae* type b bacteraemia and meningitis by intranasal inoculation.
1977	Large trial of *H. influenzae* type b capsular polysaccharide in Finland (by Helena Makela and colleagues) shows that it does not protect against meningitis in children younger than 18 months.
1978	Albert Lasker Clinical Award to Austrian, Gotschlich and Heidelberger for pioneering the development of capsular polysaccharide vaccines against the pneumococcus and meningococcus.
1980	Demonstration of improved immunogenicity of *H. influenzae* b polysaccharide by chemical coupling (conjugation) to protein by John Robbins and Rachel Schneerson.
1981	Development of meningococcal capsular polysaccharide conjugates by Harold Jennings.
1982	Porter Anderson and David Smith show that *H. influenzae* type b conjugates induce protective antibodies in human infants.
1983	Licensure of plain polysaccharide *H. influenzae* type b and pneumococcal vaccines. David Smith founds the biotech company Praxis Biologicals.
1984	First trials of *H. influenzae* type b conjugates in children.
1987	Black and Shinefield conduct large trial of *H. influenzae* type b conjugate vaccine (HbOC) showing its safety and effectiveness.
1987–88	Licensure of four distinct commercially produced *H. influenzae* type b conjugate vaccines.

1989	Use of Outer Membrane Vesicles (OMVs) as a vaccine by Finlay Institute, Cuba.
1990	Norway uses OMVs to stem outbreak of Meningococcus B invasive disease.
1991–92	Clinical trials (led by Eskola, Santosham and Booy) show efficacy of *H. influenzae* type b conjugate vaccines (PRP-D, PRP-OMP and PRP-T respectively) in infants; Takaela and colleagues show *Hi*-b conjugate vaccines reduce oropharyngeal carriage.
1995	First complete genome sequence of a free-living organism (*H. influenzae*).
1996	Albert Lasker Clinical Award to Anderson, Robbins, Smith and Schneerson for development of *H. influenzae* type b conjugate vaccines.
1999	Introduction of meningococcal C (MenC) conjugate vaccine as a routine immunisation in UK.
2000	First use of genome sequence to develop a meningococcal vaccine. Rino Rappuoli coins the term *reverse vaccinology.* Licensure of 7-valent pneumococcal conjugate vaccines.
2004	Implementation of OMVs in New Zealand to combat epidemic of invasive Meningococcal B invasive disease.
2009	Licensure of quadrivalent (A, C, Y and W) meningococcal conjugate vaccines.
2010	Introduction of MenAfriVac for prevention of meningococcal A disease in Africa through the Meningitis Vaccine Project. Licensure of 13 valent pneumococcal conjugate vaccines.
2013–15	Licensure of MenB vaccines: Bexsero (Glaxo-Smith-Kline) and Trumenba (Pfizer) in Europe and USA.
2015	Bexsero introduced into routine UK infant immunisation programme.
2020	World Health Organisation assembly approves Global Road Map to Defeat Meningitis 2030.

Prologue

The idea of writing this book first came to me when attending an Oxford college dinner in the autumn of 2015 at which I was asked to say a few words about my research. I had about three minutes to summarise 17 years of work on a new *vaccine*[a] against a devastating disease called *meningitis*. Although most of my colleagues knew this was a serious brain infection, there was confusion about what kinds of germs caused it. An expert on child education told me she had been involved in studies on children with brain damage caused by meningitis, but had no idea that it could be prevented by immunisation. An atmospheric physicist and a lead investigator of numerous planetary missions wanted to know why it took so long to develop a vaccine. A historian was concerned that immunisation could be harmful and asked whether a vaccine could cause meningitis if something went wrong. It brought home to me that even among this very educated group, so many key facts were not known or were lost in translation.

I began my medical research career in 1971 when a new era in developing vaccines against bacterial meningitis had just started. I have worked alongside and known all the major protagonists — a diverse, engaging and sometimes controversial cast of scientists. As a paediatrician, I have personal experience of looking after children with bacterial meningitis and the devastation it causes to families, their friends and their communities. But none of these experiences prepares one adequately to write a book on such a complex topic that is accessible to a wide readership. Promoting a wider understanding of science and its role in society involves that hardest of goals: simplifying without distortion. I do not subscribe to the prevalent notion that writing about science

---

[a] I have tried to minimise technical details, avoid jargon and use explanatory footnotes. Terms that may not be familiar are in *bold italics* and appear in a Glossary at the end of the book.

is too difficult or uninteresting to engage those who are curious. An author may not communicate clearly, but that is not the fault of the reader. I have tried to minimise technical details, avoid jargon and use explanatory footnotes.

Meningitis is a terrifying disease, as I know from personal experience. In the early 1990s, I was the consultant paediatrician on-call at the John Radcliffe Hospital in Oxford when a 3-year-old girl, Julia,[b] with high fever was admitted. The previous evening, Julia had lost her appetite and had spent most of the night awake, agitated and vomiting. By morning, she was lethargic and confused, complaining of a sore tummy and refusing food and drink. The admitting doctor thought it was gastroenteritis, although she seemed unusually drowsy and unwell considering her relatively brief illness. I noticed red pinpoint spots on her upper chest and neck. These alarmed me, especially when I carefully but firmly pressed her skin and the red colour didn't blanch the way most rashes do, the result of blood leaking from the small vessels into the skin. I recognised the ominous signs of the most likely diagnosis: life-threatening blood poisoning (*sepsis*) caused by a meningitis bacterium.[c] Julia was transferred to the intensive care unit. As a paediatrician specialising in infections of children and a researcher who had already spent more than a decade in the laboratory studying the steps that allow bacterial sepsis to culminate in meningitis, I understood better than most how bad the odds were that she would survive. Julia had stopped speaking or making eye contact with her parents, an indication of how quickly the infection was progressing. The bacteria had already caused serious damage to her small blood vessels, resulting in reduced blood flow and impaired delivery of essential substances (most importantly glucose and oxygen) to vital organs, such as her kidneys, heart and brain.

My involvement with such an emergency meant that it was some considerable time before I was able to talk to her shocked and panicked parents. All doctors know what it's like to be the person who must convey the terrible news that nobody wants to hear. People turn to us to provide answers, to heal and to cure. Yet, in a situation like Julia's, we're at the mercy of forces beyond

---

[b] Not her real name.

[c] These haemorrhages are typical of sepsis and meningitis caused by the meningococcus (*Neisseria meningitidis*) bacterium (see Chapter 3). These bacteria can multiply so rapidly that death may occur from overwhelming sepsis (blood poisoning).

our control. We try to do everything that medical science has taught us, but there are limits to what can be done. Within an hour of her admission, Julia had already deteriorated from being fully conscious to comatose. Her vital signs — pulse rate, blood pressure and breathing rate — were worsening. Within hours, despite all the efforts of highly trained medical professionals, Julia was dead.

The title of my book, *Brain Fever*, is an old (now obsolete) name for meningitis.[d,1] There are few descriptions of meningitis until the nineteenth century[e] when its most serious forms became as well recognised to doctors and lay contemporaries as measles, scarlet fever or tuberculosis.

Early on in my career, chance and opportunity drew me into research on the bacteria that cause meningitis, the dangerous assassins of healthy people, especially young children. This is certainly not a memoir, but it deliberately uses my personal experiences as an "insider" to trace my involvement with the pioneering research and scientists that brought about a milestone in public health: vaccines that prevent the major forms of bacterial meningitis.

Immunisation continues to ignite controversy, but it can be justifiably claimed that it is one of the greatest success stories of modern medicine. This is no trivial assertion and scientists, especially the medical profession, have a special responsibility — one we've been failing to shoulder adequately for several years — to help people understand the importance of immunisation and why it's so essential for the maintenance of public health. If the basics of what is involved in developing a vaccine and how it works are not understood, trust in immunisation is compromised. The extensive and rapid changes that our societies have undergone in the last century have left many people uncertain about the future, feeling adrift and mistrusting experts or authority figures whom they believe have misled or even lied to them. The impact of social media has been transformational, misinformation often taking precedence over reliable facts backed by evidence. It's why I wanted to write a book about one of the most feared of all diseases and explain what bacterial meningitis

---

[d] Brain fever also appears frequently in Victorian literature to describe a different malady in which fictitious characters, typically following traumatic life experiences, often died from a lengthy illness. Madame Bovary became ill with brain fever after breaking up with her lover Rodolphe. In *Great Expectations*, Pip becomes seriously ill after his father figure Magwitch dies and in Emily Bronte's *Wuthering Heights*, Heathcliff's lover also succumbs to it.

[e] The first use of term meningitis was in 1828 by John Abercrombie, a Scottish physician.

is, what causes it, who gets it and how it can be prevented. However, this is only one form of brain *inflammation* and in the later chapters I discuss how viruses, such as measles, influenza and, most recently, COVID-19, result in other forms of brain fever, including *encephalitis*. My key message is that for so many of these different diseases, the deaths and disabilities caused by damage to the brain and nervous system can be completely prevented by vaccines.

Reference

[1] Audrey C. Peterson. Brain Fever in Nineteenth-Century Literature: Fact and Fiction. *Victorian Studies*, 1976, 19(4): 445–464, Indiana University Press.

Meningitis and Vaccines: An Introduction

If you cannot — in the long run — tell everyone what you have been doing, your doing has been worthless.

Erwin Schrödinger[1]

During my years as a medical student, trainee and ultimately a professor specialising in infections of childhood, caring for children with meningitis was among the most anxiety-provoking and challenging experiences of my professional life. When it first begins, meningitis is often no different from many

Figure 1.1 Brain fever, a real and fictional Victorian disease.

1

other illnesses, perhaps seeming no worse than a mild case of "flu." But as it progresses, fever becomes pronounced accompanied by headache, cold hands and feet, vomiting and aching muscles. Breathing becomes rapid and there may be neck stiffness, dislike of bright light, seizures, altered behaviour and sleepiness. As with my patient Julia, a rash that does not fade may be an ominous warning sign. But these symptoms may happen in any order and some may not happen at all, especially in very young babies in whom recognising meningitis is notoriously difficult.

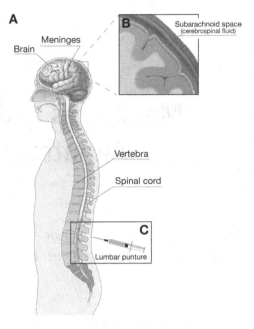

Figure 1.2　See main text for explanation.

The brain and spinal cord are surrounded by a protective triple layered covering, called the meninges (Figure 1.2 Panel A). Between the two innermost of three layers, there is a thin space containing the *cerebrospinal fluid* or CSF (Figure 1.2 Panel B) and it is within this space that the inflammation of meningitis is most pronounced. When a doctor suspects meningitis, a *lumbar puncture* (or *spinal tap*) is performed. A fine needle is inserted between the vertebrae about two-thirds of the way down the spine[a] (Figure 1.2 Panel C). Normally, CSF is clear and colourless, but when there is meningitis it is cloudy because white blood cells migrate into the fluid. This is a process called inflammation, the body's response to sensing danger when germs, such as bacteria, invade parts of the body that

---

[a] This is a routine procedure, although learning to do it expertly requires careful training. Every medical student remembers the anxiety of their first supervised lumbar puncture and one never entirely loses the fear that one may not do it successfully. Failures do happen, fortunately not often and serious complications following the procedure are very rare. However, a lumbar puncture is not always done when meningitis is suspected, especially when there are signs of raised pressure within the brain due to severe inflammation. Under these and some other circumstances, the procedure is contraindicated because it may cause serious harm.

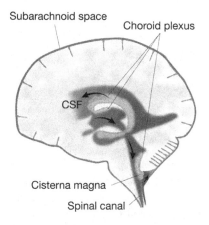

Subarachnoid space

Choroid plexus

CSF

Cisterna magna

Spinal canal

Figure 1.3   Cerebrospinal fluid (CSF) surrounds and circulates (see arrows) around the brain and spinal cord providing a protective fluid layer. CSF is produced by the choroid plexus and enters the subarachnoid space before draining into the blood stream through small porous protrusions that connect with the venous system. A rich network of capillaries supplies blood to the choroid plexus so that CSF is continuously replenished.

are normally germ-free. All of which prompts some key questions. How do bacteria get to the meninges in the first place and what happens then?

The key to understanding what is going on is to know how the body makes CSF, the protective fluid that surrounds the brain and spinal cord, provides nutrients to the cells of our nervous system and gets rid of unwanted waste materials. CSF is produced by the *choroid plexus* (Figure 1.3), a bundle of thin blood vessels and other special cells (called *ependyma*). The gaps between the blood vessels and the adjacent ependymal cells of the brain are joined closely together (called *tight junctions*) to form a barrier that blocks invasion by harmful germs, toxins and other damaging substances. The choroid plexus, richly supplied by blood vessels, produces CSF by filtering out all blood cells (red and white), larger *proteins* and other substances so that what is left is a watery liquid containing salts, smaller proteins and glucose. CSF is produced continuously by the choroid plexus, and percolates throughout the *subarachnoid space* (Figure 1.2 Panel B) before draining back into the blood stream. This cycle repeats itself so that CSF is continuously replaced. Despite the barrier created by the tight junctions, invasion of bacteria into the CSF does occur (I describe how and why in Chapter 6). The body responds by releasing special kinds of inflammatory cells (called *neutrophils*[b]) to get rid

---

[b] Also called polymorphonuclear cells or PMNs (for short).

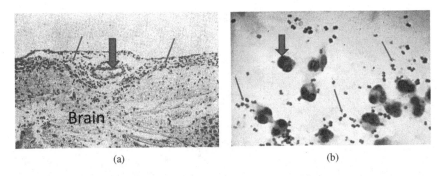

(a)   (b)

Figures 1.4 (a) Section of a brain to show the inflammation (massive accumulation of white blood cells called neutrophils) within the cerebrospinal fluid and meninges (thin arrows). A section of a blood vessel, also surrounded by inflammation, is marked by the thick arrow. (b) A sample of cerebrospinal fluid from a case of meningococcal meningitis. There are large numbers of bacteria (thin arrows) and neutrophils (thick arrow).

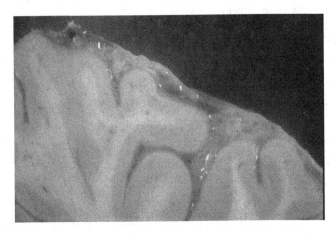

Figure 1.5 A section of a brain from a fatal case of meningitis showing the inflamed meninges forming a thick coating of pus (dead tissue and neutrophils) enveloping the *cerebral cortex* of the brain.

of the invading bacteria in what is an otherwise "privileged" site that has few mechanisms to eliminate them. But the bacteria that cause meningitis have a protective envelope on their surface (discussed in later chapters) that prevents their elimination by neutrophils. Therefore the bacteria thrive, deriving energy from the same nutrients in the CSF that are required to sustain human tissues. Each bacterium divides into two so their numbers double every 30 to 40 minutes. It only takes a few hours before there are millions of bacteria, an outcome that is associated with rampant inflammation of the CSF and adjacent meninges — meningitis (Figure 1.4).

Bacterial meningitis occurs most often in the very young but can strike at any age. Given its rapid progression, the infection is usually fatal unless treatment is given promptly. Although antibiotics are lifesaving, even these powerful drugs fail to prevent a fatal outcome in around 5%. Among those who survive, around 10% suffer from lifelong brain damage such as deafness, impaired vision, paralysis and diminished mental functioning. As eloquently summarised by Lewis Thomas:

> "… it is our response to their presence that make the disease. Our arsenals for fighting off bacteria are so powerful, and involve so many different defence mechanisms, that we are in more danger from them than from the invaders. We live in the midst of explosive devices; we are mined."[2]

How does the inflammatory mayhem in the CSF and meninges have such a drastic effect on the brain itself? Although the brain is only about 2% of the total body weight of humans, it receives about 20% of the body's blood supply. Brain damage in meningitis occurs mostly, but not exclusively, through inflammation on the blood vessels that traverse the CSF in the subarachnoid space as they enter and exit the brain. Different kinds of white blood cells, including neutrophils, surround and invade the walls of blood vessels, reducing blood supply to the brain and disrupting normal clotting mechanisms leading to thrombosis. Inflammation also damages the integrity of blood vessels, causing them to leak fluid, resulting in a build-up of excess fluid, brain *oedema*. Because the brain is encased by the rigid bones of the skull, swelling reduces the overall blood flow to the brain and collapses the blood vessels, another mechanism that interferes with its blood supply. The increased pressure can also force the base of the brain downwards, driving it into the narrow spinal canal, putting pressure on the vital parts of the brain that control breathing and consciousness, a potentially lethal event. Of course, depending on many factors, the longer-term nature and extent of the injuries to the brain caused by meningitis are extremely varied. Some are reversible, maybe completely so, especially if the diagnosis is made early and antibiotics are given promptly. But, in many instances, the inflammatory process causes irreparable damage and a tragic legacy of permanent disabilities. It should also be recognised that meningitis occurs most often in young children whose brains during the first

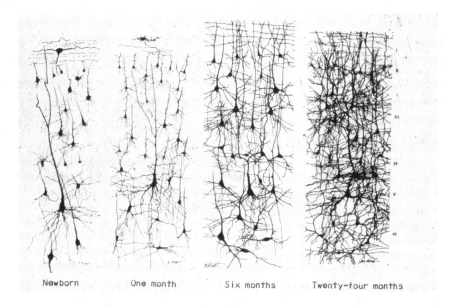

Newborn   One month   Six months   Twenty-four months

Figure 1.6 Age is a very important factor in the outcome of bacterial meningitis. Most cases occur in young children at a time when their brain is developing very rapidly. This camera lucida reconstructs the neural connections (complexity of neurons and dendritic connections) from birth to 2 years. This critical phase in brain development coincides with the most common age of occurrence of bacterial meningitis.

two years of life are undergoing enormous changes affecting the complexity and multiplicity of brain cell connectivity (Figure 1.6). This process not only imposes damage to the component nerve cells but may also interfere with the brain's programming in a way that irreversibly disrupts normal development.

Inflammation of the CSF and meninges is the classic form of bacterial meningitis, a process that occurs gradually over the hours, days and weeks following infection. But, more rarely, the same bacterial *species* that cause meningitis can result in a more abrupt illness, largely or wholly through the effects of blood poisoning (bacteraemia) and the associated life-threatening illness called sepsis. It is especially characteristic of the meningococcus. When these bacteria enter the blood, their unchecked multiplication results in widespread damage to blood vessels throughout the body so that they become leaky. The amount of blood is reduced in volume and its altered distribution impairs the delivery of vital substances, such as oxygen and glucose, to all organs — including lungs, kidneys, liver, heart and brain tissues.

This is what happened to my patient Julia, who died because of irreversible damage from sepsis before there had been sufficient time for classic meningitis to develop. But, with intensive care and antibiotics, death from sepsis is not inevitable. However, circulatory failure can have catastrophic long-term consequences, as happened with one youngster whose mother agreed to share their experience with me, capturing the enormity of what can happen after recovery from meningococcal sepsis.

When her young boy, K, became suddenly ill, it was the unusual rash that was so alarming. It prompted an internet search where she found images that were almost identical to K's rash. These were caused by multiple, small, leaks of blood (haemorrhages) into the skin and K was rushed as an emergency to the hospital. The worsening of meningococcal sepsis occurs within hours and the only way to save his life was for him to undergo amputation below the knee on one side and through the knee on the other. He also lost his left hand and all fingers of his right hand and spent weeks in hospital before he returned home to embark on the long journey of adapting to an utterly transformed life. Now a teenager, thanks to the immense support of his family, the National Health Service and the *Meningitis Research Foundation (MRF)* charity, he participates in most school activities. But anger and frustration often get the better of him and he remains very self-conscious; he told me that he just wants to be like other kids — to go swimming, to enjoy school outings — but everything is a struggle; his dependency on prosthetics is energy-sapping and he tires quickly. To purchase a bionic arm, K's parents had to obtain thousands of pounds through local fundraising. Not surprisingly, the family has suffered enormous mental and emotional stress, especially since K's baby sister was born just a week before he became ill. Growing up with an elder brother who has suffered so much has affected her deeply. Working with the MRF charity, K's parents are now strong advocates who help other parents of meningitis victims who so desperately need help and advice in facing the challenges of the life-changing disabilities caused by meningitis and sepsis.

No wonder public health surveys[3] show that meningitis is at the top of the list of infections for which a vaccine is needed. Worldwide, it affects more than 2.8 million people each year.[4] A case occurs every few minutes, resulting in one in a hundred of all child deaths, about 300,000 deaths per year,[c] although it is not just children who get meningitis.[d] But, the good news is that all the major forms of bacterial meningitis (and associated sepsis) can now be prevented by vaccines.

What is a vaccine? As early as 430 BCE, Thucydides reported during the plague of Athens that those who recovered were never affected twice and could tend to the sick without fear for their own safety. This process is called *immunity*. As early as the ninth century, the Chinese practiced *variolation*, initially by grinding up *smallpox* scabs to make a powder that was taken like snuff. By the sixteenth century, variolation was a common practice in which material from pustules of a person with smallpox was scratched into the skin of those who had not had the disease. It caused illness, sometimes fatal, but most got completely better and were protected. Edward Jenner (eighteenth century), observing that milkmaids were immune from smallpox, took fluid from *cowpox*[e] lesions on a milkmaid's hands and inoculated a young boy.[f] This seminal contribution is now recognised because Jenner was one of the first to provide written evidence — based on observation, hypothesis and experiment — of successful immunisation. Deliberate exposure to a relatively harmless or dead version of a germ activates the immune system so that it will recognise and eliminate that germ rapidly if it is encountered again.

It took several more decades before germ theory (described in Chapter 2) clarified the steps that underpin the research and development of a vaccine.

---

[c] These figures are being continually revised because of the impact of vaccines. Nevertheless, despite the availability of vaccines, there were still an estimated 5 million new cases and 290,000 deaths globally from meningitis in 2017.

[d] As a comparison, malaria causes around 600,000 deaths each year, about the same number that die from breast cancer.

[e] Vaccine is derived from the Latin *vacca* meaning cow. Edward Jenner insisted that the origin of the term *vaccination* be credited to his friend and physician, Richard Dunning, although Louis Pasteur coined the term in honour of Jenner. Cowpox causes much milder disease than the highly contagious and often deadly smallpox virus.

[f] James Phipps, the son of Jenner's gardener, who aged 8 years received more than 20 injections of Jenner's smallpox vaccine.

First comes recognition[g] that there is an infectious disease whose impact is a public health threat to society. The next step is to determine the *epidemiology* of the disease. Epidemiology is the discipline that deals with the systematic collection of data to provide details of the disease profile — the "who," "when" and "where" of the infection and how it changes over time. During this process of data and analysis, there will also be systematic research to identify the causative germ and investigate it in the laboratory. For the clinician, going through this process is an essential part of learning how best to identify and treat the disease. For the vaccine expert, the aim is to identify harmless versions of the germ or one or more of its components that can induce protection against infection. At this stage, further development depends heavily on the involvement of large pharmaceutical companies[5] to develop and produce enough vaccine to carry out clinical trials in humans. The provision of adequate evidence of vaccine safety and effectiveness is the major reason for the very lengthy time (10–15 years)[h] it takes before a vaccine can be licensed by regulators (e.g. Food and Drug Administration (FDA) or European Medicines Agency (EMA)).[i] If the vaccine proves to be safe and protective, then millions of doses will be required to carry out immunisation programmes at national and international levels within the context of the hugely demanding safety *surveillance* that is rightly demanded by governments on behalf of society.

So, how did I become involved in vaccine research? Some four years after qualifying in medicine in the UK in 1966, I became a junior doctor at the Children's Hospital in Boston, USA. There I became involved in a project to develop a vaccine to prevent one of the major forms of bacterial meningitis. But, before I embark on the story of how scientists turned the tables on the

---

[g] Different diseases can be *recognised* because there is a typical constellation of symptoms (what a person complains of) and signs (abnormal findings observed — usually by a health professional) that define a disease. As for any disease, the characteristics of those caused by microbes are such that each has similarities that are denied to others and therefore allow recognition of a discrete illness. Think of the distinct differences between, say, whooping cough and measles. Doctors make a diagnosis on much the same empirical basis as we can confidently distinguish between a dog and a cat. The sum of the parts is unique.

[h] There are exceptions; in the case of epidemics or pandemics, such as the recent Ebola or COVID-19 viruses (see Chapter 22), vaccine development may be fast-tracked.

[i] Following Brexit, the UK is now excluded from the EMA regulatory procedures which are now allocated to the UK Medicines and Healthcare Products Regulatory Agency (MHRA).

bacteria that cause meningitis, I want you to know something of the history of germs; how their role in causing serious diseases was discovered and how this knowledge resulted in the development of many vaccines, including those that prevent bacterial meningitis.

References

[1] Schrödinger, E. *Science and Humanism. Physics in our time.* Cambridge University Press Cambridge, United Kingdom, 1951. Original appears in English on pp. 8 and 9. ES gave four public lectures at the Dublin Institute for Advanced Studies at University College Dublin in February 1950 under the title: *Science as a Constituent of Humanism.*

[2] Thomas, L. *Lives of a Cell. Notes of a Biology Watcher.* Viking Press, New York, United States. 1974, 75–80.

[3] Department of Health. Childhood immunization tracking. 2011, https://www.researchgate.net/publication/261214441_Every_parents_worst_nightmare_parent%27s_attitudes_to_meningitis_and_vaccination.

[4] Zunt, J.R., Kassebaum, N.J., Blake, N. *et al.* Global, regional and national burden of meningitis 1990–2016. *Lancet Neurology,* 2018, 17:1061–1082.

[5] The Pharmaceutical Industry and Global Health. Facts and Figures. https://www.ifpma.org/wp-content/uploads/2016/01/2011.

Chapter

2

A Bacterial World

Bacteria survive, thrive, fight and die by the trillion every moment. As scientists discover more about these tiny organisms, it is becoming clear that ... we really are living in a bacterial world.[1]

In our fight against the infective diseases we are not confronted with blind forces acting at random but with the disciplined offensive of highly trained foes.[2]

The most important forms of meningitis are caused by germs called bacteria. Like so many scientific facts, this simple statement turns out to be complicated. Not all germs are bacteria; there are — in order of size (smallest to largest) — *viruses, bacteria, fungi* and *parasites*. Collectively, these different germs are often called microbes — extremely small life forms that can only be seen with a microscope. They are widespread in nature and most are beneficial to life. But a minority (called *pathogens*) are harmful to us because they can cause disease. Over the next two chapters, I'll tell you about bacteria in

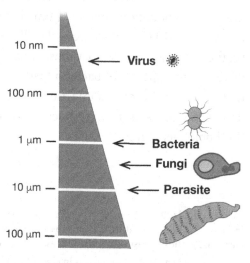

Figure 2.1   Size of microbes.

11

general and then about the different kinds (species) of bacteria that cause meningitis.

A handful of soil contains various minerals, decomposed plant or animal matter, water and some trapped gases. But, despite its lifeless appearance, it also contains more bacteria than there are people on the planet.[a,b,3] Invisible to the naked eye, there are millions of different species of bacteria whose appearance on planet Earth occurred more than 3.8 billion years ago. Each bacterium is a small, single cell, the simplest organisms we think of as being alive. Fewer than 100 species are *pathogenic* (cause disease). Together these Microbes make up around 13% of the Earth's biomass.[c,4]

Around 2.5 billion years ago, something remarkable happened. Cyanobacteria living in the oceans, commonly called blue-green algae (hence the cyan-prefix), evolved the ability to use the Sun's energy to provide their nutrition from just carbon dioxide and water, a process called *photosynthesis*. This produced oxygen that changed the make-up of the seas and atmosphere and set the scene for complex life to develop. Bacteria make about half of the oxygen that humans breathe, so we owe much of our evolution and existence to them, although without us most bacteria would do just fine. Bacteria are extraordinarily resilient, for example, *Deinococcus radiodurans* can withstand exposure to radiation a thousand times more powerful than the amount that would kill a human. Bacteria purify our water and remove much of our pollution; they, not humans, were largely responsible for cleaning up the 2010 Gulf of Mexico oil spill.[5] Although they are literally vital to us, they are also, on rare occasions, our enemy. Although most do not cause disease, a few species of bacteria, pathogens, possess the genetic make-up that can result in serious infections, such as food poisoning, pneumonia, dysentery, abscesses, urinary tract infections and, crucial in the context of this book, meningitis.

---

[a] World population is almost 8 billion (8×10^9). An average person consists of 10,000 billion human cells (10^{13}). The ratio of bacteria to each cell is estimated to be at least 1:1, possibly higher (see Ref. 3). The biomass of all microbes associated with our bodies is called the *microbiome*.

[b] A pint of typical sea water contains in excess of six million bacteria.

[c] Plants make up about 80%, compared to humans who contribute just 1/10,000 of Earth's biomass (see Ref. 4).

For most of the 2–300,000 years[6] since the emergence of *Homo sapiens* bacterial pathogens weren't so troublesome to humans. While living in small nomadic groups consisting of only a few hundred people or less, our survival depended on dodging death caused by rival tribes, wild animals, accidents, childbirth and scavenging enough food to live long enough to have children. Infections of wounds and dental sepsis were not uncommon, but deaths from communicable diseases were rare. Around 12,000 years ago, some human beings stopped being nomadic and settled in defined geographical locations. They began farming and with a sufficiency of food were able to live in permanent dwellings very near domesticated animals. This communal living was a critical factor in the evolution and spread of infectious agents that cause diseases.[d]

Today, our complex global societies and their economic drivers have resulted in cities densely populated with millions, often tens of millions, of inhabitants. The exploitation and pollution of the environment, airline travel, complex commercial food chains and animal-derived infections have made outbreaks, epidemics, and pandemics much more likely. For example, outbreaks of meningitis have been linked to large gatherings, such as the Hajj pilgrimage,[7] during which fatal cases of the disease occur as well as spread to others when participants return to their home countries.

Microbiologists have found the bacterium that causes tuberculosis (TB) in 5,400-year-old mummified Egyptians[8] and in exhumed bodies in Peru[9] dating from 1,000 years ago — or to be precise, they have been able to recover *DNA*[e] from the bacterial cells that cause TB. They have also literally unearthed DNA from bacteria found in medieval burial sites in France and England that were the cause of the Black Death,[10] the pandemic that ravaged Europe in the

---

[d] Proximity to animals who can be the source of dangerous pathogens, such as bacteria and viruses, is nothing new but remains one of the most important threats leading to pandemics, such as influenza and the 2020 COVID-19 pandemic.

[e] The cells of all life forms contain DNA (**d**esoxyrib**o**nucleic acid). DNA is made up of components (called nucleotides) of which there are four different kinds represented by the letters G, A, T, C (for guanine, adenine, thymine and cytosine). The order of these four nucleotides is a code, read in triplets of nucleotides, that specifies the building blocks (amino acids) to make different proteins. The entirety of the DNA within a cell, called its *genome*, contains thousands of *genes* that possess the information for making the proteins that carry out all the different biological functions of a cell. In later chapters, I will have more to say about the bacterial genes that are required to cause meningitis.

fourteenth century. DNA usually degrades quickly when cells die, so sampling ancient archaeological sites to find out more about the history of infections is usually uninformative.

We mostly rely on written descriptions of diseases. Writing[f] was invented around 5,000 years ago, so this places limitations on how far back we can reliably source evidence of diseases. Descriptions of illnesses that could be bacterial meningitis show up in the works of Hippocrates (460–370 BCE), the first-century Greek physician Galen (129–217 AD),[g] but most documentation[11] relies on accounts no earlier than the seventeenth century.[h] It was not until the nineteenth century that the conjunction of two concepts transformed the understanding of infection and the practice of medicine in what is called the "germ revolution."

The first was that of *contagion*, an ancient idea that in 1546 assumed a more precise identity … contagion passes from one person to another and is precisely similar in both the carrier and the receiver. The term infection is more correctly used when infection originates in very small imperceptible particles.[12] But what were these particles? This required a second discovery whose origin can be traced to the textile shop of Antonie van Leeuwenhoek in Delft, around 1650. Leeuwenhoek's favourite hobby was grinding lenses from small glass beads that were used to magnify the fabric of textiles so that clothiers could gauge the quality of their weave. But this extraordinary Dutchman was curious to understand more about the life around him and so he began using his powerful lenses to look at the fine structure of fleas, lice, flecks of his skin and hair.

---

[f] Writing — a system of graphic marks representing the units of a specific language — was invented independently in the Near East, China and Mesoamerica. The cuneiform script, created in Mesopotamia, present-day Iraq, ca. 3200 BC, was first.

[g] The Arab physician Abu Ali al-Husayn (980–1037 AD) also described "inflammation of the envelopes of the brain."

[h] Aelius Galenus (Galen) described an acute febrile condition (he called it *phrenitis*) which involved agitation and delirium associated with pus on the brain that he proposed was caused by an excess of yellow bile or a deficiency of animal spirit. By 1840, there was wide recognition of a disease named meningitis. It superseded a disparate grouping of diseases such as cephalitis, brain fever, dropsy of the brain and acute hydrocephalus. The early major contributions came from pathologists in the Edinburgh, Paris and Geneva faculties of medicine, the centres of excellence in the second half of the eighteenth century.

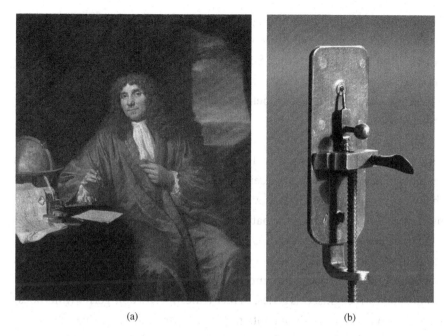

(a)   (b)

Figure 2.2   (a) Portrait of Antonie van Leeuwenhoek (1632–1723) by Jan Verkolje. (b) A replica of a microscope by van Leeuwenhoek.

Leeuwenhoek spent years perfecting techniques for grinding glass to make more and more powerful lenses with which to examine the detailed structures of tiny objects. He clamped his specimens within exquisitely fashioned metal plates, mounted only a few millimetres from his lenses. Eventually, he achieved magnification of about 500 times life-size. On his famous "day of days," he examined a drop of rainwater and saw tiny creatures, moving around with "… swiftness, as we see a top turn around, the circumference they make being no bigger than that of a fine grain of sand." He was the first person to see bacteria and what a momentous game-changing discovery it was. Up to this point, if something wasn't visible to the naked eye, it was effectively non-existent.[i]

---

[i] Van Leeuwenhoek's vision-extending discovery of microbes coincided in time and place with the vision-transforming use of the *camera obscura* by the artist Johannes Vermeer. It is almost impossible to imagine that these exact contemporaries, both baptised in 1632 and both high achievers in their fields, would not have come across each other in the small city of Delft. Indeed, Van Leeuwenhoek later served as the executor of Vermeer's estate and it has often been suggested, but not substantiated, that Van Leeuwenhoek served as the model for Vermeer's paintings *The Astronomer* and *The Geographer*. Vermeer's paintings can in part be thought of as the

Leeuwenhoek reported his findings to the Royal Society in London in 1673, claiming that in one small drop of water there were more than two million of his "little beasts." The Royal Society duly sent a delegation to Delft who verified his observations. Microbial life had been discovered, although it would take a further 200 years before the acceptance of the "germ theory" of infectious diseases.

In the nineteenth century, several seminal ideas were converging to displace the myths, miasmas and mumbo jumbo of the Middle Ages. This revolution in scientific thinking centred on understanding the role of particles that were invisible to the naked eye. Transcending the limitations of unaided human vision transformed the natural sciences. Science was rooted on structures that were invisible.[13] John Dalton's atoms (1808) became a foundation stone of physics and chemistry. The treatises of Gregor Mendel and Charles Darwin on heredity, variation and natural selection were anchored in what would later be recognised as genes. In medicine, microscopic germs (microbes) were shown to be the basis of contagion, the cause of diseases that physicians had recognised and described for centuries.

In 1845, it was discovered that airborne microbes resulted in putrefaction that caused meat to rot and wine to spoil.[14] Similar mechanisms, it was reasoned, could be responsible for the damage to human tissues caused by infections. Evidence in support of this germ theory of disease accumulated despite opposition from sceptics. Hospitals, renowned for terrible infections, such as gangrene following surgery, were a case in point. Handwashing drastically reduced the number of women dying after childbirth[j] from post-partum *(puerperal) sepsis.*

---

artistic analogue to Van Leeuwenhoek's pioneering microscopy. Through optical instruments, both allowed us to visualise what could not be perceived by the naked eye.

[j] A practice introduced in 1847 by Ignaz Semmelweis (1818–1865), a Hungarian physician appointed to the Vienna Maternity Hospital. There was a threefold increase in mothers dying from puerperal sepsis who were cared for by medical students who were going straight from the autopsy room to a maternity ward when compared to another maternity ward where mothers were attended by midwifery students. Semmelweis became known as the "saviour of mothers," although he was not popular with hospital authorities who hated to admit that they had been the unintentional cause of so many deaths. Sadly, Semmelweis soon began to exhibit what was possibly the early onset of Alzheimer's disease. He was committed to an asylum for the insane and, abandoned by his wife and friends, was beaten by the staff and died from his injuries. It

In the 1850s, Florence Nightingale insisted on making hospitals free from refuse by discarding bloodied swabs and filthy bedclothes. In 1855, John Snow famously proved that a cholera outbreak in London was caused by ingestion of contaminated water from a communal pump.[15] In the 1860s, the British surgeon, Joseph Lister, introduced carbolic acid to sterilise surgical instruments, the first widely used antiseptic. But the idea that contagion was caused by microscopic germs was not accepted by many. The miasma theory[k,16] — that diseases were caused by foul-smelling air — made sense to the sanitary reformers. Rapid industrialisation and urbanisation had created many poor, filthy and polluted city neighbourhoods that tended to be the focal points of disease and epidemics. Germs were not visible to the naked eye and few were skilled at using the microscope. By improving the housing, sanitation and general cleanliness of these existing areas, levels of disease fell. There seemed to be no need to abandon the miasma theory.

The stage was set for two of the most famous names in the history of medicine. Louis Pasteur and Robert Koch provided the crucial evidence for a concept that we now take for granted: bacteria (and other microbes) *cause* diseases.[17] This was the generalisation. The subtlety was that diseases such as puerperal sepsis and *anthrax* were caused by distinct species of bacteria. Spurred on by bitter rivalry both Pasteur and Koch were each determined to be credited with winning the race to prove the validity of the germ theory. Both made brilliant contributions. Pasteur had already discovered that certain kinds of bacteria were responsible for *fermentation*[l] of lactic acid

---

was not until 1870 that Louis Pasteur identified the bacterial cause (Group A streptococcus) of puerperal sepsis.

[k] The miasma theory proposed that diseases were caused by the presence in the air of a poisonous vapour in which were suspended particles of decaying, foul-smelling matter. In the first century BC, the Roman author and architect Vitruvius described the effects of miasma emanating from fetid swamplands (see Ref. 16).

> "For when the morning breezes blow toward the town at sunrise, if they bring with them mist from marshes and, mingled with the mist, the poisonous breath of creatures of the marshes to be wafted into the bodies of the inhabitants, they will make the site unhealthy."

Miasma theory remained popular throughout the Middle Ages and endured until the late nineteenth century.

[l] In 1859, concerning the process of fermentation, Pasteur noted that, "... everything indicates that contagious diseases owe their existence to similar causes."

(a) (b)

Figure 2.3   (a) Louis Pasteur (1822–1895). (b) Heinrich Hermann Robert Koch (1843–1910).

in milk. Between 1865 and 1870, it was shown that microbial infection of the eggs of silkworms was the cause of the shrivelling disease, *pébrine*,[m] which had devastated this prosperous French industry. These findings prompted investigations on whether the disease anthrax might be caused by microbes. It was the scourge of farmers as it would strike suddenly and kill sheep that had no contact with other anthracic animals. Contagion or miasma theory couldn't explain it. An important clue was that when scientists examined the blood of diseased sheep under the microscope, they saw stick-shaped little rods. But although suggestive, this did not provide convincing evidence that they were the cause of anthrax. It was entirely possible that something else in the blood was responsible.

Meantime, while working with anthrax in his own laboratory, Pasteur noted that old cultures of the bacteria became less potent in causing disease.

---

[m] Pasteur thought initially that it was an inherited condition but finally conceded that a rival scientist, Antoine Béchamp, had correctly discovered that it was an infection caused by a parasite. Pasteur never publicly acknowledged the importance of Béchamp's influence in establishing the importance of microbes in causing disease.

Although a simple observation, it took a genius to seize the moment. Pasteur recalled Edward Jenner's pioneering work in which he used the milder cowpox virus as a vaccine to prevent disease from the more virulent smallpox. So, Pasteur exploited the idea of using the weakened, or to use the technical term, *attenuated* anthrax bacteria as a vaccine. If successful, protection of cattle would be of huge veterinary and economic importance. But Pasteur also reasoned that if immunisation of cattle with anthrax organisms prevented the disease, it would also provide compelling evidence in support of the germ theory.

In 1882, on a small farm in Pouilly-le-Fort, France, 25 sheep were immunised with attenuated anthrax vaccine and then deliberately given large numbers of highly virulent anthrax bacteria. All the immunised animals were fully protected against disease. In contrast, 25 unvaccinated sheep died from the lethal anthrax infection. Here was powerful evidence that a specific bacterium was the cause of a disease. Although Pasteur did not fully understand the immunisation mechanism, later research showed that his vaccine was effective because it induced proteins, called *antibodies*, that neutralised the potentially lethal bacterial toxins. Pasteur's achievement brought him huge prestige.

This iconic scientist has attracted numerous biographies (many of them, interestingly and importantly, aiming to inspire the young). Yet, the mental processes and creativity that lead to discoveries in science are poorly understood. What sort of man was Louis Pasteur and what lessons can be learned about the scientific mind and how it achieves greatness? For many years (2006–2014), I was privileged to be on the Scientific Council of *Institut Pasteur*.[n] I never tired of visiting the museum located in the Institute's spacious Parisian grounds. Although most visitors' attention is taken up with the role of bacteria in fermentation and his pioneering research on anthrax and rabies vaccines, Pasteur was a chemist by training and his early work, before

---

[n] Founded in 1887, *Institut Pasteur* is world-renowned for the extraordinary contributions to microbiology and infectious diseases that were made after Louis Pasteur's death. This includes seminal research on diphtheria and antitoxins, plague, BCG vaccine for TB, phagocytes, polio virus, *bacteriophages*, antibodies and complements, typhus, *yellow fever* vaccine, sulphonamides, gene regulation and HIV. Pasteurian scientists have won ten Nobel Prizes.

he switched his attention to biology, was on crystals of tartaric acid. It was to this less frequented exhibit that I found myself drawn time and again.

Salts of tartaric acid were a problem in the wine industry because of its tendency to accumulate on the walls of wine vats. Put to work to find out more about these crystals, Pasteur found out that they existed in two spatial forms (called *isomers*), just as your hand in front of a glass mirror can be superimposed on its reflection.° This was an example of a fundamental scientific discovery driven by an industrial problem, one that had been completely missed by his illustrious predecessors and teachers in pure chemistry. The discovery changed chemistry forever. But Pasteur's research on the isomers did not stop there. He did experiments to show that they had different properties, including their contribution to the sweetness of some fruits such as grapes and apricots. This showed that subtle details in chemical structure were important to biological function. Such elegant, beautiful science — and perhaps an insight into why Pasteur went on to pursue a career that would transform medicine.

An illuminating insight into the life of this iconic scientist comes from examination of Pasteur's one hundred and two laboratory books analysed in-depth by the Princeton scientist, Gerald L. Geison.[18] I first heard about these books when chatting to a colleague during the coffee break at one of the *Institut Pasteur*'s Scientific Council meetings when he teasingly asked me if I realised that the *Institut Pasteur* was founded on deception! Geison's book provides evidence that Pasteur lied about his research, stole ideas from competitors and perpetrated what would today be considered scientific misconduct — breaching ethical principles and misrepresenting experimental data. In fairness, Geison unambiguously acknowledges Pasteur's exceptional skills as an experimentalist and emphasises how courageous he was in carrying out research on humans in an era when it would have been easy for him (a chemist with no medical training) to have avoided the risk of humiliating

---

° The different isomeric forms of the crystals could be distinguished by their ability to rotate polarised light. Pasteur separated the isomers using a magnifying lens and a pair of tweezers. The practical importance of Pasteur's discovery is well illustrated by the thalidomide disaster in the 1950s. There were two isomers of this drug, one of which caused the birth defects that affected more than 10,000 children. The other "mirror image" isomer had the sedative and anti-sickness properties that resulted in it being used for the "morning sickness" of pregnancy. Unfortunately, it was only after thalidomide had been widely prescribed that it became known that the marketed drug was a 50:50 mixture of the two isomers.

failure and public censure. But Pasteur did misrepresent the way in which he developed his anthrax vaccine in his famous trial, using a technique for weakening the bacteria that had been copied from one of his rival colleagues without crediting him.

Perhaps it is good for everyone to understand that science is not as scrupulously honest as it is often presented to be. Rhetoric, self-promotion, calculated opportunism and even cutting corners to beat rivals is prevalent in science. Using rose-tinted glasses that hold human beings to unrealistic expectations may not help the public to prepare themselves for the inevitability of scientific errors and even misconduct.

Returning to the story of germ theory, riled by his arch-enemy's coup and consumed with jealousy, Robert Koch protested that Pasteur had contributed nothing new to science, arguing that the anthrax cultures were most probably contaminated and that without techniques to guarantee purity, all bets were off. He had a point; detractors of the germ theory agreed that more evidence was needed to settle the issue.

Koch the pedagogue, sensing that his moment had come, set out the criteria needed for experiments to establish the germ theory. For each known disease, the bacterium must be isolated from the infected host and then cultured in the laboratory. That was not all; the pure culture must then be put into a healthy susceptible animal, cause the disease and then allow the causative bacterium to be recovered. Proof of disease causation rests on a concordance of scientific evidence, and what are now known as Koch's postulates[P,19-21] served as a gold standard for providing this evidence.

Here was Prussian hard-nosed rigour pitted against the Gallic flair of Pasteur. Koch now did some brilliant experiments to back up his concepts. Using boiled potatoes as a growth medium, he cultivated individual bacteria,

---

[P] These were stringent criteria to be met in establishing causation; failure to meet them in no way precluded the role of a microbe in causing disease. The power of Koch's criteria was the scientific rigour that it inspired in the early days of establishing the germ theory.

These conceptual principles underpinning causation of microbial diseases were first described by Jakob Henle in 1840 (see Ref. 19), ideas that were further developed by Robert Koch who was his student, and should more properly be called the Henle–Koch postulates.

For a scholarly analysis of their strengths and weaknesses and the later refinements emphasising the importance of epidemiological and host factors, the review by Alfred S. Evans (see Ref. 20) is recommended. In Chapter 10, I discuss how, a century later, Stanley Falkow re-interpreted these principles in the era of molecular *genetics* (see Ref. 21).

invisible to the naked eye, until they grew into a biomass of droplet sized colonies, each consisting of billions of bacteria. These were seen on the vegetable surface after a few days of incubation. Even Pasteur grudgingly acknowledged his rival's achievement. Koch's criteria could now be applied using pure cultures of bacteria, including anthrax. Between them, with contributions from their numerous scientific colleagues and assistants, they forged evidence for the germ theory. Among the discoveries of this golden era were the bacterial causes of meningitis.

References

[1] Bacterial World. An Exhibition Staged at Oxford University Museum of Natural History, October 19 2018–May 28 2019. www.oum.ox.ac.uk.

[2] Garrod, A.E. *Inborn Factors of Disease*. First published in 1931. Facsimile, Charles S. Scriver (ed.), Oxford University Press, 1989.

[3] Sender, R., Fuchs, S., and Milo, R. Revised estimates for the number of human and bacterial cells in the body. *PLOS Biology*, 2016, 14(8):e1002533.

[4] Bar-On, Y.M., Phillips, R., and Milo, R. 2018. The biomass distribution on Earth. PNAS. 115.6506-6511.

[5] Biello, D. How microbes helped clean BP's oil spill. *Scientific American*, 2015.

[6] Richter, D. *et al*. The age of the hominin fossils from Jebel Irhoud, Morocco, and the origins of the Middle Stone Age. *Nature*, 2017, 546:292–296.

[7] Saber, Y. The threat of meningococcal disease during the Hajj and Umrah mass gatherings: A comprehensive review. *Travel Medicine and Infectious Disease*, 2018, 24:51–58.

[8] Crubezy, E. *et al*. Identification of *Mycobacterium* DNA in an Egyptian Pott's disease of 5400 years old. *Comptes Rendus de l Academie des Sciences. Serie III, Sciences de la Vie (Paris)*, 1998, 321:941–951.

[9] Salo, W.L. *et al*. Identification of *Mycobacterium tuberculosis* DNA in a pre-Columbian Peruvian Mummy. *Proceedings of the National Academy of Science*, 1994, 91:2019–2094.

[10] Bos, K.I. *et al*. Eighteenth-century *Yersinia pestis* genomes reveal the long-term persistence of an historical plague focus. *eLife*, 2016. https://doi.org/10.7554/eLife.12994.001.

[11] Tyler, K.L. Chapter 28. A History of Bacterial Meningitis. *Handbook of Clinical Neurology*, 2009, 95:417–433, Elsevier.

[12] Translated by Wright, W.C. 1930 from the original paper by Girolamo Fracastoro in 1546. The translated paper appears in Part II (The Germ Theory of Disease.) pp. 69–75 in *Milestones in Microbiology*. Thomas D. Brock (ed.), Prentice Hall International, 1961.

[13] Nurse, P. *What is Life? Understanding Biology in Five Steps.* "The Cell." pp. 7–20. David Fickling Books.

[14] Pasteur, L. Memoire sur la fermentation alcoolique. *Annales de Chimie et de Physique*, 1860, 58:323–426.

[15] Steven, J. *The Ghost Map. The Story of London's Most Terrifying Epidemic — And How it Changed Science, Cities, and the Modern World.* Riverhead, 2006.

[16] Vitruvius. *De architectura.* Ten books on architecture. Ingrid Rowland (ed.), Cambridge University Press, 1999.

[17] Waller, J. *The Discovery of the Germ. Twenty Years that Transformed the Way We Think About Disease,* Columbia University Press, 2002.

[18] Geison, G.L. *The Private Science of Louis Pasteur.* Princeton University Press, 1995.

[19] Henle, J. *On "Miasmata and Cobtagie"* (translated by George Rosen). Johns Hopkins Press, Baltimore, 1938.

[20] Evans, A.S. Causation and disease: The Henle–Koch postulates revisited, *Yale Journal of Biology and Medicine*, 1976, 49:175–195.

[21] Falkow, S. Molecular Koch's postulates applied to bacterial pathogenicity — A personal recollection. *Nature Reviews Microbiology*, 2004, 2:67–72.

Chapter

3

Bacterial Assassins

Books often have heroes and villains, but few have as their central characters tiny single cells, roughly 1.0 μm[a] long and 0.3 μm wide with confusing Latin names: although many different species of bacteria can cause *acute* (sudden onset) meningitis, three are predominant and are the centrepiece of this book. Their names are a mouthful: *Streptococcus pneumoniae, Neisseria meningitidis* and *Haemophilus influenzae*. (These are species names, the same nomenclature used to refer to humans as *Homo sapiens*). I'll use the shorter, alternative names for these bacteria: pneumococcus, meningococcus and *H. influenzae* (*Hi*).

Each was identified in the last quarter of the nineteenth century, the Golden Era of Bacteriology[b] that I described in the previous chapter. The first of the triad, pneumococcus, was discovered in 1881 by an adventurous army physician, George Sternberg, when he injected his own saliva beneath the skin of a rabbit.[1] This was a strange experiment whose elusive rationale may never be completely understood. Perhaps retreating into his laboratory tempered

---

[a] This is smaller than the width of a human hair. More than a thousand lined up abreast could fit within the full stop at the end of this sentence of text. Most (but not all) bacteria are NOT visible to the human eye. The microscopic size of bacteria is important, allowing them to live in diverse niches. For example, in intertidal marine sediments, and hydrothermal vents where they thrive on nutrients that are inaccessible to other larger life forms. The small size of bacteria is beneficial, allowing them to live as parasites in a range of hosts, large and small: mammals, fish, birds, plants and insects, etc. There are limits on how big or small bacteria can get. If less than a certain size, there would not be enough space within the bacterial cell to accommodate its DNA and essential proteins. If greater than a certain size, the energy demands of the bacterial cell would be excessive.

[b] The period between 1850 and 1915 that established Germ Theory was marked by the discovery of the bacterial causes of many infectious diseases and the award of the 1901 and 1905 Nobel Prizes in Physiology or Medicine to Emil von Behring and Robert Koch, respectively.

the horrors he had experienced in the American Civil War. Taken prisoner by the Southern forces, he staged a daring escape, contracted yellow fever and went on to do battle with the Cheyenne Indians. Sternberg's rabbit died and several other animals inoculated with his spittle met the same fate. Using his microscope to examine a filtrate of his saliva, he observed chains of bacteria. In an inspirational innovation, he captured these small bacteria on camera, the first microphotograph of them ever made. He grew the bacteria in bouillon broth and demonstrated their pathogenic role. The freshly cultured bacteria were lethal when inoculated into rabbits, thus fulfilling one of Robert Koch's key criteria for establishing the role of a bacterium in causing an infectious disease. Months later, Sternberg's findings were confirmed independently by Pasteur. Both noted that the bacteria were surrounded by an aureole or halo of some unknown substance. These newly discovered bacteria were shown to be the cause of pneumonia, *septicaemia* and meningitis in humans. I'll return to this fascinating microbe and the mysteries of its halo — and why it became known as the "sugar-coated microbe" — in Chapter 5.

In 1805, the Swiss physician Gaspard Vieusseux reported what is likely to have been the first clear description of an outbreak of meningitis.[c,2] It happened in Eaux-Vives,[d] a precinct located outside the city walls of Geneva in Switzerland. Today it's an up-market neighbourhood on Lake Geneva, with a five-star hotel overlooking a landmark fountain that sends a jet of water more than 400 feet high. It's hard to imagine Eaux-Vives's wretched conditions at the turn of the nineteenth century when today Switzerland's average income is the equivalent of €50,000. But in the brutal winter of 1805, this village was home to about 900 desperately poor labourers and their families, who were crowded into primitive dwellings. They survived on a hunk of bread during the day and a bowl of potato soup at night with a cube or two of meat on Sundays. A dreadful stench from the lake, a public latrine where Geneva dumped its sewage, added further indignity. According to Vieusseux's written reports, an astonishingly rapid and severe febrile illness affected the people of Eaux-Vives before spreading to the richer precincts of Geneva. At the end of January 1805, two of the three young children of a single mother died within 24 hours of contracting an

---

[c] He called it "*fièvre cérébrale maligne non contagieuse*" (non-contagious malignant cerebral fever).

[d] Literally "Running Waters."

illness that was characterised by high fever, severe
headache, rigidity of the
neck, vomiting, delirium,
loss of consciousness,
convulsions and purple
patches on the skin due
to haemorrhages. By the
time Vieusseux arrived,
the first of the children
was near death, and the
doctor felt helpless as
the boy's pulse became
feeble, then undetectable,
and he died only 12 hours
into the illness.

It was the rapidity
and severity that imme-
diately marked the condi-
tion as different from the
numerous other fevers
that were relatively com-

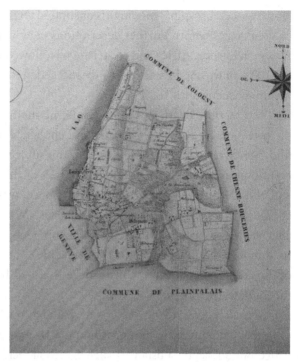

Figure 3.1   Eaux-Vives largely consisted of allotments
adjacent to scattered primitive wooden huts and primitive
dwellings mostly close to the lake and just adjacent to the
boundary separating the commune from the city.[e]

mon at that time, especially among the poor and malnourished families that
lived in crowded conditions close to the lake. When Vieusseux and a local
pathologist colleague[3] performed an autopsy, they found a thick coating of pus
overlaying the base and back of the brain, the result of inflammation of the
linings of the brain, the hallmark of meningitis. Reports of a similar disease in
Medford, Massachusetts, USA, occurred just a year after the Geneva epidemic,
although the two American doctors[f,4-6] who described this outbreak, aptly

[e] I am grateful to the staff of the Geneva State Archives for their help in obtaining authentic
documents and a map of *Eaux-Vives* from the early nineteenth century.

[f] Lothario Danielson and Elias Mann. They described nine children with a dramatic disease in
which livid spots appeared under the skin, on the face, neck and extremities. There was fever,
vomiting, and rapid progression to coma. They called the disease cerebrospinal meningitis or
the spotted fever (see Refs. 4, 5). Their original paper described: "a singular and very mortal
disease characterised by violent pain in the head and stomach succeeded by cold chills."

known as cerebrospinal meningitis or spotted fever, were apparently unaware of Vieusseux's report. Further descriptions of epidemics in North and South America, Europe, Africa and western Asia followed where deaths occurred in 70–100% of the victims. But nobody succeeded in observing or growing the meningococcus until 1887.[g,7]

The terrible death toll from epidemic meningococcal disease placed the infection on a par with plague and cholera. Huge efforts were made to find effective treatments, opinions on the subject leading to heated, often acrimonious discussions among physicians of the day. In this era, emetics and bloodletting were still in vogue, but opium was the favoured remedy of choice for meningococcal meningitis because of its stimulant properties. To alleviate the unbearable headaches of meningococcal meningitis, physicians used repeated lumbar punctures with the idea of reducing the high pressure of the cerebrospinal fluid, a procedure that had been widely used to treat tuberculous meningitis.

The ready access to the cerebrospinal fluid through the improved techniques of lumbar puncture led to the first attempts at treatment through the spinal canal[h] with crude antiseptic substances, such as carbolic acid, but these were ineffective or harmful. A breakthrough in treatment came from pioneering research showing that *serum* from the blood of horses that had been immunised with meningococci could protect laboratory animals against meningococcal disease. This was a two-step process called *passive immunisation*.[i] First, animals were injected with meningococci and then, after there had been enough time for them to develop immunity, the animals were bled. From the blood, the serum was extracted and used as treatment. Simon Flexner, a pioneer of this approach, showed that injection of this immune serum into the spinal canal halved the number of deaths

---

[g] The introduction of lumbar punctures as a routine clinical procedure cemented meningococcus as a major cause of meningitis. Under the microscope, they were typically arranged in pairs, often inside pus cells, and were therefore originally called *diplococcus intracellularis meningitidis*. This was later changed to *Neisseria meningitidis* (1901), named after Albert Neisser, who discovered the closely related bacterium *Neisseria gonorrhoea* (1879). Anton Weichselbaum was the first to grow the elusive and fastidious bacteria from the spinal fluid (see Ref. 7).

[h] The so-called intrathecal route in which drugs or treatments were injected into the fluid-filled space between of the meninges (between the pia and the arachnoid). See Figure 1.1, Chapter 1.

[i] Passive immunisation was pioneered in 1890 by Emil von Behring and Shibasaburō Kitasato and later recognised by the 1901 Nobel Prize.

from meningococcal meningitis.[8] It was a big stride forwards, but only partially successful. Flexner did not understand why his serum treatment protected against some cases of meningococcal infection but not others, an important piece of the jigsaw that would be solved by his colleagues at the Rockefeller Institute as I describe in Chapter 5.

The discovery of the final member of our triptych of meningitis-causing bacteria, *Haemophilus influenzae* (*Hi*)[j], came about through the kind of circuitous and non-obvious series of events that is typical of the way science unfolds. In the winter of 1889–1890, an epidemic of influenza occurred in Russia. It spread worldwide resulting in the first modern influenza pandemic. From the respiratory secretions of individuals dying from influenza, Richard Pfeiffer isolated microscopic stubby, rod-like bacteria that he proposed were the cause of influenza. A former student of Robert Koch, he worked as his assistant in the Institute of Hygiene in Berlin (1887–1891) and it seems likely that these bacteria would have been the subject of discussions between them.[k] The assertion that *Hi* bacteria were responsible for influenza brought Pfeiffer huge prestige, especially when in 1918–1919, there occurred an even more devastating influenza pandemic (the infamous *Spanish flu*), a global catastrophe that was far more serious than the *Russian flu* some 20 years earlier.

Much has been written about the effects of the 1918 influenza pandemic in which more people died (the upper limit of the death toll of influenza has been estimated at 8–10% of all young adults) between September and

---

[j] The origins of *Haemophilus* stem from its growth requirements in the laboratory; it has an absolute requirement (*philus* = love) for factors derived from red blood cells (*haem* = blood). In medieval times, *influenza* (Italian, but familiar in English as *influence*) was used to describe the many epidemic illnesses that were thought to be the result of occult visitations from the heavens or supernatural forces. Thus, in the vernacular, *Haemophilus influenzae* might be liberally translated as "the blood loving germ of evil influence."

[k] In fact, Robert Koch had discovered the same bacterium when he went to Egypt to investigate the widespread occurrence of epidemic conjunctivitis (often known as pink eye). He used microscopy to examine the purulent discharge of infected eyes and saw stubby rods. Another scientist, (John Elmer Weeks) made similar findings and established their causal role in acute conjunctivitis using the criteria proposed by Koch. He inoculated the secretions from several members of a family who were suffering from conjunctivitis into the eyes of six human volunteers, five of whom developed disease. Pus from each of these infected persons showed the typical stubby microscopic bacteria that he was able to grow in pure culture.

December of 1918 than died on the battlefields during the Great War.[1] Influenza killed more people in a year than did the Black Death in a century and more than AIDS has done in four decades. As the enormity of the influenza pandemic escalated throughout 1918, the inadequacy of the evidence of a role for *Hi* (also known as Pfeiffer's bacillus) in causing influenza changed from a minor controversy to a *cause célèbre* of international importance. The credibility of the scientific establishment from whom governments, health services and the public required immediate and authoritative information was on the line. Embarrassingly, it was clear to most that the correctness of Pfeiffer's conclusions, ultimately shown to be erroneous, owed more to the imperious reputation of bacteriologists in the early twentieth century than to scientific objectivity. Experimental infection of laboratory animals did not make them ill, so the postulates of Pfeiffer's mentor, Robert Koch, had not been fulfilled. Epidemiological methods, such as case control studies, were not then part of the repertoire of scientific investigation. The proposed role of *Hi* in causing influenza was based only on an association. Crucially, it was not appreciated that most healthy people are *carriers* of one or more *Hi* strains so its recovery from healthy, as well as sick, individuals was to be expected.

Society hates an impostor and the mistaken role of *Hi* in causing influenza might have relegated the bacterium to obscurity. But its legacy was assured when it was shown to be the major cause of childhood meningitis. Ironically, the first authentic case of *Hi* meningitis had been described in 1899, Pfeiffer himself supervising the bacteriologic work. During the years that followed a few scattered cases were reported, but it was not until 1911 that a real impetus was given to the study of *Hi* meningitis by a New York pathologist (Martha Wollstein) who described a series of eight cases of fatal meningitis in children aged 5 months to 4 years. This was a classic study from a remarkable medical scientist.[9] Appalled by the tragic deaths of such young children, she made a special study of the *Hi* bacteria that had been cultured from their spinal

---

[1] The total number of military and civilian deaths from World War 1 was around 40 million. The Spanish flu killed in excess of 50 million. The Black Death killed an estimated 25 million. (However, it is important to know that the world population around the year 1350 was only 370 million. Proportionately, the Black Death caused about double the number of deaths as did the Spanish flu.) In 2020, there have been 33 million deaths from HIV since the beginning of the pandemic.

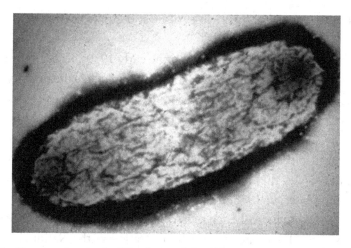

Figure 3.2   Electron micrograph of *H. influenzae* b (*Hi*-b) bacterium. The capsule on the outside of the bacterial cell is stained densely black. N.B. The same photo but with some additional information and a structure appears in Chapter 6.

fluids. To investigate their *virulence*, she injected them into goats and found that they caused much more serious illness than did the *Hi* bacteria that had been isolated from cases of influenza. To find out why, she did a whole series of investigations. The results were confusing and, with hindsight, it's easy to understand why. At the time, like many of her colleagues, Wollstein wrongly believed that *Hi* caused influenza and that meningitis was one of its serious complications.[m] But influenza and *Hi* meningitis were distinct diseases,

---

[m] Some interesting perspectives on how slowly the research on *Hi* became widely accepted in clinical practice is evident from textbooks of the time. Osler's Practice of Medicine (3rd Edition), published in 1899, considered epidemic influenza to be caused by *Hi* and there is no mention of it as a cause of sepsis or meningitis of *infants*. The situation was still confused in 1920. In the description of epidemic influenza, Holt's 8th edition of *Diseases of Infancy and Childhood* opined that: "… the correctness of Pfeiffer's views (is) … questioned by many good observers." Holt's 9th edition (1926) cites *Hi* as one of the secondary causes of meningitis and considered it to arise as one of the manifestations of a more generalised form of the disease. Not until 1935 does Osler's *Practice of Medicine* (13th edition) state that there is a type of meningitis caused by *Hi*. By 1935 (13th edition) Osler's text states that *Hi* meningitis has nothing to do with the disease usually spoken of as influenza. No new edition of Holt's textbook appeared between 1933 and 1940. The text of the 11th edition contains the classic chapter contributed by the paediatrician Dr Hattie Alexander of New York's Columbia University. It provides a comprehensive summary of *Hi* meningitis, the role of capsular *polysaccharide* and lists the other important diseases caused by the bacterium.

although it was not until the early 1930s that there was definitive proof that influenza was caused by a virus.[n,10,11]

At the Rockefeller Institute, also in the early 1930s, research by another remarkable woman scientist, Margaret Pittman, unlocked the mystery of what made the *Hi* bacteria such a deadly cause of meningitis. She discovered that there was a protective coating or capsule (see Figure 3.2) on the surface of the *Hi* bacteria[12] that was essential for the bacteria to cause meningitis — a finding that was inspired by the prior discovery of the sugar coating surrounding the pneumococcus, the "halo" that I mentioned earlier in this chapter. More than 40 years later, in 1975, I had the privilege to meet this gracious, nononsense *grande dame*. Although retired and in her mid-seventies, she worked most days at the Food and Drug Administration at the National Institutes of Health (NIH) in Bethesda, where pioneering work on a vaccine against *Hi* meningitis was being driven along at full throttle. I was a young paediatrician, based at Johns Hopkins Medical School, doing research on meningitis, and a frequent visitor to the NIH. But I am getting ahead of my story; here is how all this came about.

References

[1] Sternberg, G.M. A fatal form of septicaemia in the rabbit, produced by the subcutaneous injection of human saliva. *Annual Reports of the National Board of Health*, 1881, 2:781–783.

[2] Vieusseux, G. Mémoire sur la maladie qui a régné à Genève au printemps de 1805. *J. Med. Chir. Pharmacol.*, 1805, 11:163–182.

[3] Matthey, A. Recherches sur une maladie particulière qui a régné à Genève en 1805. *J. Med. Chir. Pharmacol.*, 1806, 11:243–253.

[4] Adams, D. (ed). Reviews of Infectious Diseases: *The First American Account of Cerebrospinal Meningitis*, 1983, 5: 969–972. Extracted from *The Medical and Agricultural Register, for the Years 1806 and 1807*, Vol. 1, No. 5.

[5] Vedros, NA. *Evolution of Meningococcal Disease*. Volumes I and II. Boca Raton CRC Press, 1987.

---

[n] Richard Shope of the Rockefeller Institute led the way with his studies on swine influenza (see Ref. 10). His techniques were used by scientists at the London's National Institute for Medical Research to show that human influenza was caused by a virus (see Ref. 11).

[6] Hirsch, A. Epidemic cerebro-spinal meningitis. In *Handbook of Geographical and Historical Pathology*, Vol. 3. The New Sydenham Society, London, England, 1886.

[7] Weichselbaum, A. Uber die Aetiologie der Meningitis Cerebrospinalis Epidemica. *Fortschr. Med.*, 1887, 5:573–583.

[8] Flexner, S. Experimental cerebrospinal meningitis and its serum treatment. *JAMA* 47, 1906, 560–566.

[9] Wollstein, M. Influenzal Meningitis and its Experimental Production. *Am. J. Dis. Child.*, 1911, 1:42–58.

[10] Shope, R.E. and Francis, T. The susceptibility of swine to the virus of human influenza. *J. Exp. Med.*, 1936, 64:791–801.

[11] Smith, W., Andrewes, C.H., and Laidlaw, P.P. A virus obtained from influenza patients. *Lancet*, 1933, 222:66–68.

[12] Pittman, M. Variation and type specificity in the bacterial species *Haemophilus influenzae*. *J. Exp. Med.*, 1931, 53:471–492.

Becoming a Medical Doctor

I had wanted to be a doctor from a young age although it's hard to discern how and why this resolute commitment happened so early in my life. I recall my parents reading a book entitled *A Surgeon Remembers*, written by the prolific George Sava, the pseudonym of a British doctor who had been born in Russia. When told I was too young to understand it, of course I was determined to read it. The book consisted of a collection of puzzling medical case studies with a common theme; the answers to what had made the patient ill were not to be found in the textbooks; the doctors were baffled, then Sava intervened and the patient was saved. The notion that you could save a person's life and be held in awe by your colleagues thrilled and impressed me.

Up until my teenage years, I do not believe that I showed any vestige of academic promise other than an intense, undisciplined curiosity of a superficial kind that my teachers found aggravating (and told me so). A poor listener, I had a short attention span and was intellectually lazy and careless. Then around 14, at Shrewsbury School, I underwent a rather sudden transformation in attitude, ambition and awareness; I began keeping a diary:

Michaelmas Term 1956
Two older boys in the same study are getting up at 6 in the morning to work before breakfast. One is working on submitting a Greek poem for the school's prestigious classics prize. The other is editing the school magazine. I must do something too. I have entered for the school junior student Chemistry Prize; an essay on "The Contributions of German chemists to Physical Chemistry."

Easter Term 1957
My housemaster thinks I ought to try for a place at Cambridge to study medicine. My science teacher's doubtful as I must do organic chemistry for

the Cambridge University entrance exam. Only solution is do the mandatory Advanced level General Chemistry in one, not two years. Have been given go-ahead! My friends think I'm crazy and making a big mistake.

Summer Term 1957
My biology teacher has shown more interest in me since he learned about my doing Chemistry A level in one year. He thinks I may not do well in Biology. He didn't like my essay. I wrote that: "... the purpose of the heart is to act as a pump to distribute blood around the body ...". He told me that the heart doesn't have a purpose. He has loaned me a small book by Ernest Baldwin: An Introduction to Comparative Biochemistry with several passages underlined in pencil. Among them: "... we must not suppose that modifications ... could be or ever were purposively established." I think I understand what he is getting at. The book is amazing: I love reading it, although there are many things I don't understand. He's told me that's normal and not to worry.

Almost everyone recalls at least one teacher from school whose influence stands out and outlasts others and so it was with my biology teacher. The content of his teaching was often unusual with quirky anecdotes reflecting life events, characters from novels and subtle observations that demanded thought. It also felt as if what he had to say, whether it was telling us about Charles Darwin or the intricacies of the nervous system, was especially directed at me, as if he knew and shared my thought patterns. Although there were about 20 others in the class, he made *me* feel that I was his most important and favourite student and I did not want to disappoint him. I am still uncertain as to what it was that gave me that sense of special rapport, but it inspired me. There was something enigmatic and vulnerable about him; he would often seem to be lost in his own thoughts, excuse himself and set us a short writing task while he spent time in the teacher's room smoking in his inimitable manner one Passing Cloud cigarette after another.

I had some aptitude for biology and chemistry, very little for physics, but got a place at St. John's College, Cambridge to study medicine. My time at university was an erratic rollercoaster of self-examination and a painful process of coming to terms with what seemed an unending confusion of how to balance the demands of work and play. Medical students bore a much greater burden of obligatory activities than students in other disciplines with many lectures, practical sessions and tutorials. My girlfriend, a law student, introduced me to her large circle of friends who were studying languages, history and architecture; they appealed to me far more than my colleagues in medicine — not least because they were able to take a more relaxed approach to their studies. By the end of my first year, the demands of medical studies and the pleasures of social life lay in stark opposition. After ten years of privileged but constrained life in boarding schools, I was free to immerse myself in whatever experiences I wanted and take advantage of the diversity and excitement around me. I was exploring new ideas and meeting an incredible cross-section of talented people. My tutors told me in no uncertain terms that I was falling behind. I hated anatomy, the dreariness of mugging up seemingly endless details required a commitment that robbed me of precious hours for my other interests. I did not study anatomy thoroughly and failed my exams outright. I was humiliated, facing a crisis and knew that I had to pull myself together. I reduced my socialising, re-sat anatomy and was judged to know enough to graduate.

I left Cambridge for St. Thomas's Hospital in London, where I did my clinical training from 1963–1966. I was now among medics, no longer in daily contact with the diverse academic and social ambience of Cambridge University, so it was easier to dedicate my time to my medical studies. But my whole attitude had changed after the disaster of my exam failure as an undergraduate. I had learnt the hard way that there was no substitute for putting in long hours.

After qualifying, I became a junior doctor at the age of 25. My decision to specialise in paediatrics was serendipitous. I helped a friend by temporarily taking on his job on the children's ward of the Whittington Hospital in North London so he could go skiing. Out of the blue I discovered the special challenges and satisfactions of looking after children. Most sick children, even those who are seriously ill for a time, recover. Not always of course and my next job at the Hospital for Sick Children, Great Ormond Street (GOS), proved hugely demanding. I was a junior doctor responsible for children with complex illnesses that were referred to this famous hospital.

My working week exceeded 100 hours and my boss, who also worked at St. Thomas's Hospital and had a thriving private practice, only spent two half days at GOS. Indeed, being responsible for private patients was part of a house officer's job, although strictly speaking we were employed exclusively by the

National Health Service. Late one afternoon, my boss telephoned me about a referral from a London private paediatrician. A two-year old boy was due to arrive by ambulance. He had developed high fever and, according to his grandmother, was not at all himself. "It's as if someone had taken the batteries out of him" was how she described his behaviour. She was sure that he was not at all well and had something serious. On arrival, he seemed to have improved and had only a mild fever, but given the alarming history and previously high temperature, I did a lumbar puncture that to my relief went smoothly. The cerebrospinal fluid looked crystal clear and as I was preparing to send it and blood tests to the laboratory, I was greeted by the Resident Assistant Physician (RAP), an experienced, plain-spoken but hugely respected "Aussie" who as the most senior "on-call" GOS paediatrician wanted to be sure all was going OK. Bill Marshall was a legend, an enormously capable general paediatrician who would later become the first specialist in childhood infections in the UK. On his insistence, we personally took the blood and cerebrospinal fluid samples to the pathology laboratory. When not acting as RAP, Bill was doing research on rubella virus (the cause of German measles) and so knew all the laboratory staff including the on-call technician. We quickly established that the apparently clear CSF contained a few abnormal cells — a clear indication of inflammation — although not enough to make the fluid turbid. So, the young lad had meningitis (Grannie was right!). But Bill then did something else. He stained a droplet of the CSF that he had put on a glass slide before spending several minutes peering at it under the microscope. Then, in his typical pithy, disparaging Aussie style he told me to look sharp and see what he had found. He had identified the microscopic, stubby rod-shaped bacteria that were typical of *H. influenzae (Hi)*. It was an unforgettable learning experience. We had a precise diagnosis within an hour of the boy's admission.

On ward rounds the next morning, despite many hours of antibiotic treatment overnight, the boy's fever had settled but he was less alert and had considerable stiffness of his neck that had not been present the evening before. I was dismayed, deeply worried that I had not given the correct treatment. "Seen it before," my boss explained, adding "luckily his meningitis was diagnosed early, but it's not stopped the inflammation even though the bacteria have been killed by the antibiotic." It was another important lesson. Treatment of meningitis is lifesaving but does not necessarily prevent the potentially damaging inflammation to the brain caused by the body's immune responses to the bacteria. It was my first case of caring for a child with *Hi* meningitis. Fortunately, in this case, he made a complete recovery.

During my time at GOS, I was responsible for looking after many other very ill children. I had to draw on all my resources to care for a baby with an extremely rare genetically inherited condition called "Maple Syrup Disorder." She lacked a crucial *enzyme*, without which she could not break down proteins without accumulating toxic substances. The baby was only one of a few dozen cases reported in the literature and she required round-the-clock care.

This was also the era when improved treatments for childhood leukaemia were being developed and tested through complex protocols within clinical trials to find out what combination and doses of potent, cytotoxic drugs were most effective. A diagnosis of acute lymphatic leukaemia in the 1960s was virtually a death sentence; today, more than 90% of cases are cured. But this incredible progress, from relatively ineffective treatments to high rates of curing leukaemia, came about through a series of clinical trials using potent but extremely toxic drugs. Participating in the protocols was a brutal and demanding ordeal for patients, parents, nurses and doctors. The relentless pressure of my job came to a head with a child whose treatment was failing to halt the aggressive course of her leukaemia and whose parents were becoming fraught because of the appalling side effects of the treatment. I was caught in the middle and saw my role as being an advocate for the child. My frank-ness was not appreciated by my boss who took me aside later and gave me a dressing down, creating tensions between us that prevailed in the remaining weeks of my six-month job, causing me great personal distress. Close to melt down, I was befriended by a visiting young American physician who was part of an exchange programme between the Children's Hospital of Philadelphia and GOS. He convinced me that it would be useful to gain experience on the other side of the Atlantic. I duly filled in an application and, to my surprise and delight, was accepted to do a year of training at the famous Boston Children's Hospital.

After finishing at GOS, I had a few months before I was due to start in Boston. Relaxing over a pint of beer, one of my more enterprising colleagues suggested that I apply to be a ship's doctor for a few months. "It'll be an adven-ture," he enthused, "You won't have to do much; it's basically a paid holiday." That sounded like my sort of vacation, and I took myself to the P&O shipping office and made enquiries. Before long, I'd signed a three-month contract as doctor for the crew and passengers of the SS *Hardwicke Grange*, a merchant ship ploughing the sea lanes between Plymouth, and Buenos Aires.

I was excited and not a little nervous as the merchant ship slipped its moorings and the land receded as we headed out on the green phosphorescence of the open sea. Crew members were introduced to "Doc," pints of Export Watney's were downed, and I slowly accustomed myself to the pitch and sway of the boat, the chug and clank of engines. My surgery hours, less than two hours each day, were light work compared to the long hours as a junior doctor. The three years since I had qualified had been exhilarating but utterly exhausting. I now had time to think, to mull over my experiences and reflect on what my future expectations were. There had to be more to life than being on the treadmill of work experience as a junior doctor. There would be much time in the next weeks for me to explore these nascent thoughts that were shaping my aspirations concerning my future career.

Meantime, on the first morning, I dealt with four cases of severe seasickness — three among the passengers and one from a 16-year-old novice cabin boy, who came to me holding his stomach and weeping. As it turned out, the sickness he was suffering from was of the home kind rather than maritime. We stopped for a disappointingly short two hours at the Canary Islands before reaching the port of Recife in northern Brazil. Here we docked for 36 hours, and the passengers and the crew disembarked for much drinking and the pleasures of local women. Following our departure from the harbour, my surgery became busy and I needed large doses of intramuscular penicillin to treat several cases of the "clap" (*gonorrhoea*). The unfortunate homesick cabin boy, who had been initiated into the rituals of seafaring life, was initially reluctant to seek treatment. Wisely, an older member of crew stopped by the surgery and told me "That kid is in such pain, Doc, he was screaming as he was peeing."

Before reaching Buenos Aires, we docked in Montevideo, in Uruguay, where I bought tickets for the home country to play their arch-rivals, Argentina, at football. I have always loved sport, but little did I realise just how passionate South American fans can be. I thought there would be a riot when Argentina was awarded a penalty against Uruguay late in what had been a goalless game, to secure a controversial win.

Just before Christmas, we sailed down the Rio de la Plata to Buenos Aires, the buildings in the city centre still bearing the scars of the military coup three years earlier. I boarded the train that ran along the Rio de la Plata to Tigre where I enjoyed a very festive Christmas. Plenty of wine from Mendosa and a feast of various meats cooked on the traditional Argentine *parilla*.

Within hours of our departure, I faced the most serious medical challenge of the trip. One of the crew literally lost his mind and went on a rampage, threatening people and property. He was suffering from *delirium tremens*, and it took four people to wrestle him into submission and place him in a straitjacket, which allowed me to give him the maximum recommended dose of paraldehyde. After a few minutes, he fell unconscious, but I was worried about further complications, and stayed with him for several hours before, thankfully, still constrained by the straitjacket he threatened to tear everyone limb from limb. That crew member was not the only one who had a sobering experience from that voyage.

After this, the weeks passed without incident and I began to become bored and tired of the confines of life on board. I was now thinking about what lay ahead in Boston, latently energised, restless and ready to be challenged. I spent hours on deck, feeling the rhythmic undulations of the sea as our ship carved its way through the choppy Atlantic waters, wondering what awaited me in America and what opportunities it might bring. I was ready and impatient for something new, but uncertain as to what.

As we entered the port of Rotterdam in the early hours of January 22, disaster struck. It was cold and foggy, and in the gloom our ship collided with a grain elevator that was outside its navigational waters and in our shipping lane. In the poor visibility, the ship's master tried to avoid collision, but succeeded only in diverting the *Hardwicke Grange* into the side of a Soviet Cruiser on its maiden voyage, inflicting costly damage. More tragically, four members of the grain elevator crew were dispatched into the icy waters where they became hypothermic and drowned. It was a horrible finale to my South American adventure. My departure for North America could not come quickly enough.

Chapter

5

On the Shoulders of Giants

Being a young trainee in the USA in the 1970s was very different from anything I'd experienced in the UK. After medical school, the training of hospital doctors in the UK was largely through work experience;[a] the teaching was loosely structured and junior doctors were so busy that there was little time for dedicated training sessions or further study. In contrast, the residency programme (comprising around 100 trainees) at Boston Children's Hospital was highly organised with numerous lectures, journal clubs and other supervised training activities. But most striking of all were my fellow junior doctors, including bottom-of-the rung *interns* many of whom had already been immersed in major, cutting-edge research activities. Several had combined studies for their medical degrees (MDs) with a substantial research project culminating in a PhD, affectionately referred to as "mud-fuds." I recall having coffee with one as he recounted with pride how a few months earlier he'd given a talk on his research at an international cancer conference. This seemed utterly incredible to me. I was impressed and felt inferior and envious. It was a wake-up call, one of several reasons why, after only a few weeks in Boston, I made up my mind to stay in the US longer than one year. In fact, I was to remain for fourteen.

The US in the early 1970s was a thrilling place to be. The atmosphere was charged with social change — civil rights, women's rights and the impact of the Vietnam War. The unmistakable odour of marijuana pervaded Boston's Harvard Square. I was equally thrilled by the unbridled enthusiasm of my

---

[a] There have been substantial changes in the UK during the past 30–40 years with the introduction of more formal requirements (overseen by the Royal Colleges) for training of hospital specialists, public health and general practice. However, implementation of these curricula has been very variable and the heavy demands of service provision to an overstretched National Health Service still mean that work experience rather than structured training dominates the training of junior doctors.

fellow trainees. The buzz and vibrancy were such a contrast to the often morale-sapping "trench warfare" atmosphere among junior doctors in the UK. Many high-profile, inspiring senior staff members were role models, clinician-scientists who gave generously of their time in mentoring us. The leadership of Charles A. Janeway was especially prominent. He had been responsible for setting up the Boston Children's Hospital clinical research programmes, of which I would later become a part. His research interest was *immunology* — how the body defends itself against microbes; he was one of the first to describe the deficiencies of immune mechanisms in children that made them unusually susceptible to infections. He was instrumental in setting up a research programme to prevent *H. influenzae (Hi)* meningitis. The basis of the meningitis vaccines that I am going to describe in this book was largely the legacy of decades

Figure 5.1   Charles A. Janeway (1909–1981) built the first department of paediatrics in the United States that was based upon the new developments in basic sciences. He travelled widely teaching modern paediatrics to thousands of physician throughout the developing world. He played a role in founding many teaching hospitals in the Middle East and Africa. His son, "Carly" Janeway was an internationally distinguished immunologist who was the first to identify what is now known as *innate immunity.*

of research carried out at the Rockefeller Institute and University between 1920–1970. Although it is something of a diversion, it is essential as historical background to go back in time to describe how scientists laid these foundations. It's a great example of how the building blocks of science are fashioned.

The Rockefeller complex of buildings dominates the East side of the Hudson River in Manhattan, New York. Its nondescript buildings are utilitarian rather than eye-catching, the stark, prosaic brickwork now tempered by a thick covering of ivy. The interior of the main building, called Founder's Hall, features expansive marble stairways with beautifully carved oak bannisters. A magnificent painting of Antoine Lavoisier, donated by John D. Rockefeller Junior in 1927,[b,1]

---

[b] Although the money came from the Rockefeller family, I learned from Robert S. Desowitz's book (see Ref. 1) that the impetus for the philanthropy came from Frederick T. Gates, a Baptist

attests to the central importance of the chemical view of medical research that was at the heart of the Institute's vision. The origins of the Rockefeller Institute were linked to personal family tragedy. John D. Rockefeller Senior's grandson died from scarlet fever in January 1901. On the advice of close colleagues and family, an institute was founded to tackle the major bacterial infectious diseases that were at the time the greatest known threats to human health: tuberculosis, diphtheria, scarlet fever, typhoid fever, pneumonia and meningitis.

Germ theory had provided a basis for combatting, even preventing, the appalling impact of these infections. At the end of the nineteenth century, there were already legendary biomedical research centres in Europe, most prominently the Koch and Pasteur Institutes. However, at that time, until the Rockefeller Institute was opened in 1903, the United States had virtually no dedicated facilities for the investigation of infections. The Institute's mission was to link laboratory investigations to bedside medicine thereby providing a scientific basis for disease detection, prevention and treatment — an idea promoted by the legendary William Osler of Johns Hopkins Hospital in Baltimore who urged his colleagues: "See, and then research ... But see first." The Rockefeller Institute, with its laboratories adjoining a hospital, was the enactment of this vision. It became the model for the dozens of other clinical research centres established in the next decade through which the United States became the global leader in biomedical research.

Among the major influential figures of the time was William Henry Welch, the former Head of Pathology, also from Johns Hopkins. Gregarious and corpulent, he, like Oscar Wilde's Lord Darlington, "... could resist anything but temptation," especially when it came to ice cream, sweets and carnivals in Atlantic City, which he adored. Welch remained based at Johns Hopkins, but one of his protégés, Simon Flexner, became the first director of the Rockefeller Institute in 1906. He was the pioneer of a successful treatment of meningococcal meningitis (see Chapter 3). A large outbreak in New York occurred in 1904 and Flexner's public health intervention using injections of immune serum halved the mortality. Flexner's approach had been inspired by Paul Ehrlich, the father of modern immunology. In 1900, Ehrlich discovered proteins in blood that bound to the surface of bacteria. He called them antibodies, noting that

---

visionary who had also been the inspiration behind the *hookworm* campaign and the fight against yellow fever.

they neutralised microbial toxins and killed bacteria.[c] Antibodies were the active ingredient in Flexner's serum therapy, but he did not know what components on the meningococcal bacterial surface were being targeted. The denouement of this critical question came about through a group of researchers that Flexner presciently recruited to the Rockefeller Institute.

At the forefront of this research was Oswald Avery, a graduate of the College of Physicians and Surgeons in New York City and one of the greatest biomedical scientists of all time. After three years as a full-time clinician, he abandoned hospital practice to commit himself to laboratory research. From his appoint-

Figure 5.2   Simon Flexner (1863–1946). In 1899, a few years before Flexner became Head of the Rockefeller Institute, he discovered the cause of dysentery (shigellosis) while in the Philippines. His elder brother, Abraham Flexner, was responsible for the Flexner Report (1910), a landmark contribution that shaped many of the existing educational principles of contemporary training in US medical schools.

ment to the Rockefeller Institute in 1910 until his retirement in 1947, Avery's research was largely devoted to understanding the biology of the pneumococcus. In the US in the early 1900s, pneumococci were responsible for around 50,000 deaths per year from pneumonia, sepsis and meningitis. Avery was convinced that the complicated biology of infections could only be understood through adopting a chemical approach. This amalgam of biology and chemistry,

---

[c] Ehrlich's ideas on antibodies were inspired by the "lock and key" hypothesis for enzymes that had been proposed in 1894.

biochemistry,[d] was a new discipline in Avery's hey-day. Avery wanted to know why in some people the pneumococcus was so virulent while in others the infection was aborted. These were fundamental questions applicable to all serious infections. The choice of the pneumococcus turned out to be an ideal model organism in which to define the biochemical basis of bacterial virulence.

Avery's line of enquiry was inspired by an eccentric British bacteriologist, Fred Griffiths. When pneumococci were grown on a solid growth medium, he noted large and small colonies of bacteria. The larger ones seemed to exude a mucoid substance giving them a smooth glassy appearance (called S forms); the surface of the smaller colonies was lacklustre and pebbly or rough (called R forms). The S forms caused fatal infection when injected into animals, whereas the R forms

Figure 5.3 Oswald Avery (1877–1955) was a pioneering scientist who was the first to isolate DNA, the chemical basis of genes, which was arguably one of the most important discoveries in the history of medicine. This fascinating story is told in-depth in books by Maclyn McCarty (*The Transforming Principle*, 1985) and René Dubos (*The Professor, The Institute and DNA*, 1976).

completely lacked this virulence. Avery mentally pictured that the S forms of pneumococci were enveloped in a thick gel, a protective capsule, that prevented the bacteria from being removed by the body's immune mechanisms. When pneumococci were grown in flasks, some of this capsular material was released into the culture fluid.[e] More impressively, it was also found in the urine of patients suffering from pneumococcal pneumonia or meningitis. Because it dissolved in body fluids during pneumococcal infection, Avery called it

---

[d] The term was formally coined by a German chemist, Carl Neuberg, in 1903. However, its conceptual origins owe much to research in the second half of the nineteenth century, for example Claude Bernard who united the disciplines of physiology and chemistry and Louis Pasteur who recognised the importance of microbial cells in fermentation.

[e] This key observation was made by A. R. Dochez, Avery's long-standing friend and Rockefeller colleague with whom for many years he had daily scientific discussions.

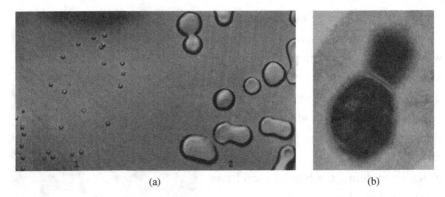

Figure 5.4 (a) The contrasting appearance, when grown on a solid growth media, of smooth (S), encapsulated pneumococci (on right) and the much smaller rough (R) capsule-deficient colonies (on left). (b) Electron micrograph showing a dividing pneumococcus bacterium surrounded by its capsular polysaccharide. It is not difficult to envisage why the "gloopy" encapsulated S forms were more shielded from the body's immune defences.

the *specific soluble substance*. Years went by and all his efforts to identify its chemical nature ended in failure. He needed the expertise of a chemist and so it was that he approached a colleague, Michael Heidelberger, to whom he brandished a small vial of brownish powder, an extract of the capsular material that he had isolated from the S forms of pneumococci. Avery was convinced that understanding its chemical nature was the secret to understanding the virulence of the pneumococcus.

In 1922, the two began work on this project. Heidelberger recalls how he went down with Avery to the basement cold room to look for some pneumococcal antiserum. All the bottles were contaminated with fungi. Avery was horrified and embarrassed, but Heidelberger was undeterred; he knew that he could use the mouldy antiserum for his chemical analyses;[f,2] sterility was not important for the method he intended to use. Both assumed that the pneumococcal soluble substance would be made of protein; the prevailing dogma at the time was that all molecules of biological importance were proteins. But Heidelberger's analyses showed that the unknown material had no nitrogen

---

[f] Providing that the fungi had not degraded the antibodies in the antisera, Heidelberger knew that what he could get would bind to the soluble specific substance in what is called a *precipitin reaction*. He could then obtain a complex of the antibody and the capsular material. This was an important step towards purifying the specific polysaccharide for further analytical tests on its chemical composition.

and therefore could not possibly be a protein. It was Avery who then suggested the unthinkable: could the soluble substance be made of sugars? The test for sugars was strongly positive, one of which was immediately identified as glucose.

The pieces of the biological puzzle were beginning to fall into place and Avery talked endlessly about his wonderful bug, "the sugar-coated microbe." Further analysis showed that the capsular material consisted of long chains of sugars joined together to form

Figure 5.5   Michael Heidelberger (1888–1991). It was his training as a chemist that got him interested in studying molecules on the surface of cells. Through Carl Landsteiner he became interested in red blood cells and the chemical basis that determined different blood groups. Changing his interest to bacteria and their surface molecules was just a different application of chemistry to a biological problem. The work with Avery underpinned the basis of today's meningitis vaccines. Here he seems to be eying the tube that Avery gave him to find out the chemical basis of the pneumococcal specific soluble substance.

polysaccharides. There were differences in the chemical composition of the polysaccharide capsules depending on which isolate of pneumococcus was analysed. This is where antibodies were crucial to their further investigations. Analogous to a lock and key, antibodies (proteins) bind to a target substance in a chemically specific manner and so could be used to distinguish the different polysaccharides. To obtain a set of antibodies, animals were immunised with different polysaccharides.[g] But there was a problem. The yields of antibodies were often very poor and hindered their progress.

Then Avery learned of a trick that solved the problem. It was the brainchild of Karl Landsteiner[h,3] who had joined the Rockefeller Institute in 1922.

---

[g] These were distinguished based on the marked differences in the surface appearance of pneumococcal bacterial colonies isolated from patients and grown on agar plates in the laboratory.

[h] The studies were carried out with James van der Scheer. For a detailed insight into their work, see Ref. 3.

Landsteiner's research was completely different; he was an expert on red blood cells and had discovered that there were different sugars on their surfaces — findings that were the basis of what we know today as the A, B and O blood groups. Indeed, this seminal discovery paved the way for giving safe blood transfusions for which Landsteiner was later awarded a Nobel Prize.[i] At one stage, just like Avery and Heidelberger, Landsteiner's research had run into a problem because he could not make antibodies to the red blood cell surface sugars and was unable to differentiate one blood group from another. The solution was to link the sugars to proteins[j,4] whereupon he found that immunising animals with these protein-sugar complexes was highly efficient in inducing antibodies.

Taking their lead from this work, Avery and Walther Goebel coupled pneumococcal capsular polysaccharides to proteins, such as egg albumin. Eventually, this resulted in highly reactive, specific antibodies to the different pneumococcal capsular polysaccharides. It was a long and demanding project — published in a series of 13 papers between 1920 and 1931.[5]

Antibodies were not only used to characterise the capsular polysaccharide antigens but were also of huge relevance to clinical practice. Further investigations showed that these antibodies protected animals against lethal pneumococcal bacterial infection. For many years they were used as the main treatment for pneumococcal pneumonia in humans — reminiscent of Simon Flexner's earlier serum therapy against meningococcal meningitis. But now Avery and colleagues had discovered the scientific rationale. The targets for the protective antibodies were distinct capsular polysaccharides. It made sense that Simon Flexner's serum therapy only worked if the antibodies targeted the correct capsular polysaccharide.

It was the *recognition* of the S forms of the pneumococcus that inspired Margaret Pittman's identification (1931) of six distinct *Hi* capsular polysaccharides (designated a, b, c, d, e and f) and her seminal observation that the b

[i] Nobel Prize for Physiology or Medicine, 1930.

[j] Landsteiner carried out these seminal studies from 1922–1943 at the Rockefeller Institute. He showed that small molecules, called *haptens*, only stimulated antibody production when combined with a larger molecule. This was later known as the hapten-carrier principle. His work was summarised in a classic text (*The Specificity of Serological Reactions*, 1936 (see Ref. 4)) that established the field of immunochemistry.

capsular type caused almost all cases of meningitis.[6] In 1933, similar findings were made for three capsular polysaccharides of meningococcus,[7] designated A, B and C — all of which were major causes of meningitis.

Decades of research could now be summarised in a sentence. Capsular polysaccharides were the major *virulence factors* of the bacteria responsible for serious bacterial infections, including meningitis, and specific antibodies were protective. Based on these findings, the chemist Michael Heidelberger and a clinician colleague, Colin MacLeod, worked together to develop the first vaccines using purified pneumococcal capsular polysaccharides and in 1943–1944, clinical trials carried out in 17,000 military recruits showed that they were highly effective.[8]

However, what followed is a salutary lesson in what happens in the practice of medicine. Shortly after the success of this breakthrough in immunisation, the widespread availability of sulphonamides and penicillin pre-empted the use of vaccines. It took more than a decade, by then the 1960s, before Robert Austrian showed that, despite antibiotic treatment, the mortality from serious pneumococcal infections was still unacceptably high. The importance of vaccines was resurrected.

Austrian spearheaded the development of a vaccine against 14 capsular types of pneumococci that was widely used. He was the first to document that the immune response of infants to capsular polysaccharides were strikingly inferior to older children and adults.[9] Many years later, I would meet and become friendly with this charming, passionate scientist and quintessential American gentleman. He told me that as an adult physician and infectious diseases specialist, he had not appreciated the profound importance of the unresponsiveness of young children when immunised with his polysaccharide vaccine. Meningitis occurs most commonly in the first years of life,[k] the very time that immune responses to polysaccharides are least effective, the implications of which will become clear in later chapters.

---

[k] According to the US Centres for Disease Control the most common causes of bacterial meningitis beyond the new-born period are Hi-b, pneumococcus and meningococcus. In children less than five years of age, these three pathogens make up about 75% of all episodes of acute bacterial meningitis worldwide.

Meanwhile, in the 1960s, coinciding with the Vietnam War, a strategic decision was made by the United States Army to invest in research into communicable diseases and preventive medicine because of the need to maintain the health and efficient performance of combat troops. Meningococcal disease, especially among new military recruits, had been recognised as far back as World War I. In 1916, the massive recruitment of young men in the UK, on any one day averaging more than 13,000, to fight on the Western Front meant that the training depot at Caterham in London housed many times the number of new recruits that could be accommodated in the barracks. Instead, the young men were housed under canvas in what was an unusually cold and wet winter. Crowding was the major factor that resulted in the escalation of meningococcal carriage and outbreaks of disease.

Outbreaks of this kind were not confined to the London barracks. In research[10] on the lives and deaths of 47 men buried in the Commonwealth war graves at her local church near Salisbury, it was discovered that the vast majority of these soldiers from the Great War, who had been housed in similarly cramped conditions in camps on Salisbury Plain (1917–1922), had expired from meningococcal meningitis — not as was popularly believed from Spanish flu (the pandemic outbreak that killed more people worldwide than those who'd died in the fighting). An army physician[11] virtually eliminated meningococcal disease in this kind of setting by increasing the available sleeping space, providing open windows, and curtailing the length of time that recruits were clustered together on parades, an early example of "social distancing."

In World War II, the armed forces of Germany, France, Norway, Denmark, Australia and New Zealand experienced major outbreaks of meningococcal disease. Antimicrobial treatment with the antibiotic sulfadiazine was used to prevent person-to-person spread of meningococci. By the 1960s, this treatment ceased to be effective because during the build-up of the US military through conscription, antimicrobial-resistant strains of meningococcus caused several highly publicised outbreaks resulting in suspension of basic training. Public concern proved to be as difficult to manage as the disease itself. It was in this context that, in 1966, the Walter Reed Army Institute of Research (WRAIR) in Washington DC responded by initiating a research programme dedicated to controlling the impact of meningococcal disease and its deleterious effects on the armed forces. The WRAIR scientists set about investigating the extreme susceptibility of army recruits to meningococcal disease. A key to the feasibility and success of the research was the strict (enforced by the military) compliance of the recruits. These researchers were therefore able to obtain nose and throat cultures as well as blood samples and then monitor the recruits over a few years to detect cases of disease in a large cohort at high-risk of meningococcal disease. The findings were conclusive. Disease occurred in those who lacked antibodies to the capsular polysaccharides of meningococcus,[l,12] a fact that encouraged the idea of a vaccine. Emil Gotschlich, a young scientist whose career had been centred at the Rockefeller Institute and who was brought up in the wake of the decades of research on capsular polysaccharides, developed the key laboratory assay used to detect antibodies to meningococcal polysaccharides in the blood of army recruits. He also purified batches of the C capsular polysaccharide[m] variant of the strains causing disease in the military. Immunisation of new recruits with the C polysaccharide vaccine resulted in virtually total prevention of disease — a major triumph in public health.[12]

The development of effective pneumococcal and meningococcal capsular polysaccharide vaccines in adults was recognised in 1978 through the prestigious Albert Lasker Clinical Medical Research Award to Michael Heidelberger, Emil Gotschlich and Robert Austrian. But what of Avery whose scientific genius inspired so much of the edifice underpinning the early generation of polysaccharide vaccines? I suspect that Avery, so precise and modest, would have been embarrassed by this assertion; but the facts speak for themselves. Indeed, in the early 1930s Avery was nominated almost yearly for the Nobel Prize for his and Michael Heidelberger's discovery that the virulence and

---

[l] These studies, for which Malcolm Artenstein and Irving Goldschneider deserve huge credit, became classics as they described a technique that measured the activity of human sera (mediated through the actions of complement and antibody proteins) to kill virulent meningococci. The results were recorded as the dilution of sera that killed at least 50% of the standard inoculum of bacteria in the assay mix. This serum *bactericidal assay* became the "gold standard" for defining a surrogate of *protective efficacy* against each of the different capsular serogroups of meningococcus. It has been internationally adopted and used by regulatory bodies as a benchmark for assessing meningococcal vaccines prior to *licensure*. It is still used today, more than 50 years after it was first used in the studies on US army recruits.

[m] Other scientists (most notably Elvin Kabat) at the Rockefeller had tried to make a meningococcal polysaccharide vaccine but were unsuccessful. The key to Gotschlich's success was that he devised a method that obtained polysaccharide preparations of high *molecular weight* that were immunogenic in humans. At the same time as he worked on the C variant polysaccharide, he prepared batches of polysaccharide from the A and B capsular meningococcal variants.

immunity of pneumococci depended on its polysaccharide capsule. But many scientists were critical of this conclusion and suggested that the antigenic properties depended on contaminating protein despite that Avery's group had compellingly dispelled this objection. Nonetheless, the Nobel committee considered that the research was not worthy of the Nobel Prize.

Although it's something of a digression, it is worth adding that in 1946 came a further stunning discovery by Avery and his colleagues. It had all started around 1927 when a British researcher, Frederick Griffith,[n] the first to describe the S and R forms of pneumococci mentioned earlier in this chapter, observed that multiple distinct strains of pneumococci, each making a different polysaccharide capsule, could be grown in culture from the sputum of individuals with pneumonia. This could be explained easily enough by proposing that individuals were often infected, through distinct transmission events, with more than one strain of pneumococcus. But, apparently, Griffith didn't favour this explanation. Ignoring the principle of parsimony, Griffiths plumped for a bolder premise. He hypothesised that somehow there'd been a "transformation," in which some of the original pneumococci had been altered resulting in their expressing a different capsular polysaccharide. His hunch was backed by evidence; when mixtures of killed pneumococci of one capsular type and live pneumococci of another type were injected into animals, some of the live pneumococci acquired the capsular polysaccharide type of the killed organisms. Interestingly, the experiments did not reject or support his hypothesis of multiple serotypes of capsules; but the discovery of *transformation* was seminal. On reading about these experiments, Avery and his research team spent 15 years researching the mechanism and chemical basis before concluding in 1944 that "the transforming factor" consisted of nucleic acids.[o,13] This discovery was the first persuasive evidence that DNA was the chemical basis of genes, arguably the most important discovery in

---

[n] Griffiths and Avery never met — Avery practically never travelled. Griffiths was killed in an air-raid in London (1941).

[o] Avery had Graves's Disease and was incapacitated enough to be out of the laboratory for much of the time when the key research on the *Transforming Principle* was done. The story of the discovery of nucleic acids as the biochemical basis of heredity is beautifully described by one of the key scientists, Maclyn McCarty (see Ref. 13).

physiology or medicine of the century. So why didn't Avery get the Nobel Prize?

The entrenched idea that genes were proteins did not die easily. Further, Avery was quiet, self-effacing and presented his work in a low-key manner. When invited to speak at larger meetings he usually sent his younger collaborators. He refused to travel to England to receive the prestigious Copley Medal from the Royal Society. Whether or not this reticence affected the Nobel committee, they concluded that evidence for DNA as the transforming principle was insufficient and that Avery did not merit a Nobel Prize. In 1953, Watson and Crick published the structure of DNA and any lingering doubts about the nature of the transforming principle were utterly dispelled. The "double helix" was indeed the molecule of life.[14] Now surely the prize was assured, but Avery died in 1955 of liver cancer and Nobel laureates cannot be recognised posthumously. Jim Watson, Francis Crick and Maurice Wilkins were awarded the Nobel Prize in Physiology and Medicine in 1962 for "… their discoveries concerning the molecular structure of nucleic acids and its significance for information transfer in living material."

The legacy of the Rockefeller Institute, the culmination of brilliant research over several decades, is a prime illustration of the slow, painstaking accumulation of essential information that is typical of great science. "Eureka moments" make for an exciting story; journalists and their readers love the stereotypical idea of the scientific breakthrough. However, as the Rockefeller story shows, this kind of sensationalism is largely a misrepresentation of how most scientific progress is achieved. There's usually no clear road map, rather the unfolding of an eclectic process, where seemingly unconnected findings are first forged into rough patterns, then refined into more coherent and ordered concepts.

So it was that the threads of disparate research over many decades converged to lay the foundation of our current meningitis vaccines. This was where matters stood in the late 1960s. There were vaccines for adults, but little was known about their effectiveness in children. But as Charles Janeway knew only too well, meningitis was predominantly a disease of very young children. Now, as the Head of Boston Children's Hospital, his championing of research to develop a vaccine against the major cause of childhood meningitis would herald a new era. It coincided with my chance to become personally involved in meningitis research.

References

[1] Desowitz, R.S. *Federal Bodysnatchers and the New Guinea Virus: Tales of Parasites, People and Politics*, p. 122, WW Norton, 2002.

[2] Heidelberger, M. A "pure" organic chemist's downward path. *Annual Review* of *Microbiology*, 1977, 31:1–13.

[3] Silverstein, A.M. "A History of Immunology," Chapter 6. The concept of immunologic specificity. 2nd Edition, Elsevier, 2009.

[4] Available in a revised edition published (1947) by Harvard University Press.

[5] Avery, O.T. and Goebel, W.F. Chemo-immunological studies on conjugated carbohydrate-proteins: 11. Immunological specificity of synthetic sugar-protein antigens. *Journal of Experimental Medicine*, 1929, 50, 533–550.

[6] Pittman, M. Variation and type specificity in the bacterial species *Hemophilus influenzae*. *Journal of Experimental Medicine*, 1931, 53:471–492.

[7] Rake, G. and Scherp, H.W. Studies on meningococcus infection. III. The antigenic complex of the meningococcus — A type-specific substance. *Journal of Experimental Medicine*, 1933, 58:341–360.

[8] MacLeod, C.M., Hodges, R.G., Heidelberger, M., and Bernhard, W.G. Prevention of pneumococcal pneumonia by immunisation with specific capsular polysaccharides. *Journal of Experimental Medicine*, 1945, 82:445–465.

[9] Austrian, R. Life with the pneumococcus. Chapter 4, p. 58. University of Pennsylvania Press, 1985.

[10] Rowe, H. Unpublished personal communication.

[11] Glover, J.A. Spacing-out in the prevention of military epidemics of cerebrospinal fever. *BMJ*, 1918, 2:509–512.

[12] Goldschneider, I., Gotschlich, E.C., and Artenstein, M.S. Human immunity to the meningococcus. I. The role of humoral antibodies. *Journal of Experimental Medicine*, 1969, 129:1307–1326.

[13] McCarty, M. *The Transforming principle: Discovering that Genes are Made of DNA*. WW Norton, 1985.

[14] Watson, J.D. *The Double Helix*, 1968. Atheneum Press (USA); Weidenfeld and Nicolson (UK).

Becoming a Clinician-Scientist

A few months into my time in Boston in 1970, I decided to seek Janeway's advice about my future. I was anxious to remain in the US where I was completely caught up in the excitement of a different lifestyle and career opportunities. I entered Janeway's spacious office where he was seated next to a bay window, his profile silhouetted by the bright light of late afternoon sunshine. It was his habit to see people while at the same time dealing with his voluminous correspondence. I don't recall much of what I said to him, but I must have rambled on about wanting to do research after my year of residency was completed. For what seemed an age, Janeway was silent. Finally, a few softly spoken words reached me: "Well, I will see what I can do. It is my impression that you are suited to do research if that is what you want."

A few weeks later, I received a telephone call to say I'd been accepted to do a research fellowship in the Boston Children's Hospital Infectious Diseases Division. It was only many years later that I learned that the money for my fellowship had come from Janeway's personal fund, one created from honoraria he'd received for his numerous distinguished lectures.

The Chief of the Infectious Diseases Division was David Smith, a tall, physically intimidating and disarmingly intelligent mid-westerner. He was later to confide in me how influential Janeway had been as his mentor, inspiring him and other young doctors to expand their vision and to use their clinical experience to identify

Figure 6.1   David Smith (1932–1999).

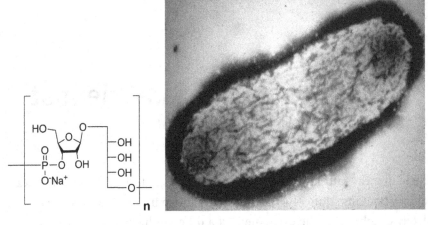

Figure 6.2   Electron micrograph of *H. influenzae* bacterium. By the early 1950s, the chemistry of the type b capsular polysaccharide was found to be a *polymer* of multiple repeats (n) of the sugar ribose and its ribosyl derivative (linked through phosphate), called (*polyribose-ribosyl phosphate*) or PRP.

important problems that needed further research. At the bedside of a child suffering the agonies of bacterial meningitis, Janeway challenged his protégé to develop a vaccine to prevent *H. influenzae (Hi)* meningitis. A seed was sown. Smith decided, with Janeway's encouragement, to take time off from clinical medicine and immerse himself in basic research. After completing a PhD, Janeway recruited Smith to head up a newly created Infectious Diseases Division, where he assembled a team of brilliant young scientists.

The *Hi*-b vaccine project was underpinned by two pieces of seminal research carried out decades previously. It will be recalled that in 1931 Margaret Pittman had shown that almost all *Hi* meningitis cases were caused by isolates making the b type capsular polysaccharide. In 1933, a review of the medical records at the Children's Hospital showed that 80% of *Hi*-b meningitis cases occurred between two months of age and six years. The Boston scientists thought that the reason why the disease was so deadly in young children was because of a lack of antibodies in their blood. To test this idea, they incubated *Hi*-b bacteria with blood samples from individuals of different ages. Blood from young babies did not kill the bacteria, whereas samples taken from children older than 5 years were lethal. Here was evidence that protective blood factors disappeared very rapidly after birth[a] following which babies were highly susceptible to meningitis until natural exposure[b] to *Hi* resulted in active immunity. It is a beautiful example of how scientific knowledge progresses through observation, hypothesis and experiment. The findings remain to this day among the most frequently cited scientific articles[1] on meningitis.

One of David Smith's new recruits to the project was Porter Anderson. Often ponderous and seemingly socially awkward, his thoughtful and penetrating intellect marked him out as an exceptional scientist. After studying chemistry as an undergraduate, he had worked in Honduras where he did research on pesticides before doing his graduate studies in the same Boston laboratory as David Smith. After completing his PhD, deeply sensitive to the troubled world around him, he felt an urge to help mitigate racial injustice and took a teaching position at Stillman College in his native Alabama, where most of the students were black. But David Smith shrewdly realised that Porter had just the right skills[c] that were needed for the *Hi* project and recruited him to Boston in 1968. It was to prove an inspired decision; Porter was crucial to the development of the *Hi*-b vaccine; he also became one of my mentors and a lifelong friend.

In 1971, I joined David's team as a trainee in infectious diseases of children.

Figure 6.3   Porter Anderson (1937–).

---

[a] Blood samples from new-born babies up until a few weeks killed the bacteria because they possessed antibodies that had crossed the placenta from the mother. But this maternally derived immunity waned sharply after a few weeks.

[b] As explained earlier, *Hi* spreads from person to person. This usually results in harmless carriage of the bacterium, even with the virulent type b variants. Invasion into the blood is a rare event, so that even in those who lack antibodies, there are hundreds more youngsters who are carriers than there are cases of disease.

[c] This capsular polysaccharide, **p**olyribosyl-**r**ibitol **p**hosphate, known as PRP (containing D-ribose and phosphodiester linkages) was then considered a derivative of the nucleic acid, *RNA*. Having done a PhD in the same laboratory as Porter Anderson, David Smith knew that Porter Anderson had experience in RNA purification, one of the reasons why he recruited him in 1968 to join the *Hi*-b project.

Inexperienced and new to research, my first assigned task was a systematic review of the current literature on bacterial meningitis. "Don't accept anything without sifting the evidence for and against any assertion" was the advice of my supervisor Arnold Smith (no relative of David Smith). Then in his mid-thirties, a graduate of the University of Missouri, Arnie's rise in the academic ranks of US paediatrics had been nothing less than spectacular. Already an assistant professor at the Boston Children's Hospital, he was renowned for his encyclopaedic knowledge and for his distinguished track record in meningitis research. He was the youngest *clinician-scientist* ever invited to be on the editorial board of the *New England Journal of Medicine*. Prodigiously energetic, he loved motorcycles, fast cars (especially Porsches), and gourmet food. I still have a classic text on Biochemistry that he gave me, in which he inscribed a quote from William Osler, the first physician-in-chief at Johns Hopkins Hospital and later the Nuffield Professor of Medicine at Oxford.

Medicine is learned by the bedside and not in the classroom ... See, and then research ... but see first.

What I saw was a chance to try and understand exactly how these bacteria strike previously healthy infants causing devastating infections. I read about the three major species of bacteria[d] that cause meningitis. Some of the main things that I learned were that the bacteria that cause meningitis can only survive in humans, not in other animals, plants, or insects, nor in soil, water or other inanimate niches. Their survival depends on living in the upper airways of humans (*colonisation*), although this so-called "carrier state" lasts for at most a few months. So, without transmission from the upper respiratory tract of one person to another, the bacteria that cause meningitis would peter out. But these bacteria do spread within communities, via respiratory droplets and secretions, so that at any given time, a proportion of persons is colonised with one or more of these three bacterial species, the number varying from a few percent to more than half of the population — depending on factors, such as age, concomitant viral infections and crowding. Most people who acquire the bacteria remain in good health. Everyone reading this book has or will become a carrier of these potentially devastating bacteria — but

---

[d] See Chapter 3.

you will be relieved to know that you're many, many thousands of times more likely to be a healthy carrier than to come down with meningitis. But there is a sting in the tail because on rare occasions, these bacteria breach the linings of the nose and throat and spread to the brain. When this biological trespass occurs, it typically does so within days or even hours after a person who lacks immunity acquires the bacteria. But why it happens in only an occasional carrier is not well understood.

Ironically, causing invasive disease ultimately isn't an advantage to the pathogen. Once the invading bacteria enter the blood stream, they cannot be transmitted to other people because the bacteria are now confined in a closed location from where there's no escape. Their fate is sealed because the bacteria will either kill the host (no advantage to that) or be killed by our natural immune mechanisms, antibiotic treatment, or both. Causing disease seems accidental or incidental to their lifestyle, conferring no survival advantage. Darwin, who barely mentions bacteria in his many books on evolutionary theory, would have been intrigued.

After a lot of reading, I was anxious to get to work in the laboratory. My goal was to develop an experimental animal model to study how meningitis happens and to evaluate the effect of vaccines on preventing meningitis in the early phases of their development — long before carrying out clinical trials. To be effective, vaccines must pre-empt one or more of the stages of the infection *before* there is injury to the brain. I discovered from my literature review that all previous animal models of meningitis had missed out the crucial initiating step. Instead of starting the infection in the nose and throat area (nasopharynx), prior animal models of meningitis had been induced by injecting the bacteria directly into the spinal canal or blood.

I decided to imitate as closely as possible the human infection by instilling *Hi* bacteria into the noses of laboratory rats[e] and then trace the sequential steps that eventually resulted in meningitis. My initial efforts to induce infection via the nasal route encountered technical problems. Arnie Smith had advised me to use baby rats (aged only a few days) because they were known to be highly susceptible to infection and easy to handle — being about the same size as an adult mouse. I found it difficult to get the droplet of the bacterial suspension

---

[e] The model used infant rats aged 5–10 days, an age at which the animals lacked immunity and were highly susceptible to meningitis.

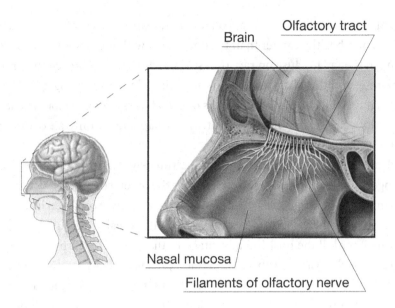

Brain

Olfactory tract

Nasal mucosa

Filaments of olfactory nerve

Figure 6.4   The nerve filaments for sensing smell ramify into the upper part of the nose. Given their proximity to where meningitis-causing bacteria are naturally found (nasopharyngeal carriage), this ascending route is evidently a likely pathway for the bacteria to cause meningitis. Although the balance of evidence does not support this idea, many experts consider that it is involved in a minority of cases of bacterial meningitis. For example, in meningococcal infection, the arid conditions and dust storms in sub-Saharan Africa may cause damage to the linings of the nasal passage, possibly accounting for some cases. Naturally occurring or accidental damage to the bony plate (ethmoid) separating the base of the brain from the nasal cavity is also well documented in some cases of pneumococcal meningitis. This more direct bacterial invasion is more difficult to prevent by immunisation than is invasion by the blood stream.

of *Hi*-b organisms into the tiny opening of the animals' noses. Using a head lens helped and with practice I could accurately deliver the bacterial inoculum in the correct location. Whether this would induce experimental meningitis was completely uncertain. As Peter Medawar, a famous British scientist and Nobel Prize laureate in physiology and medicine, says in his *Advice to a Young Scientist*: "There is no certain way of telling in advance if the day-dreams of a life dedicated to the pursuit of truth will carry a novice through the frustration of seeing experiments fail and of making the dismaying discovery that some of one's ideas are groundless."[2]

Theories at the time proposed that *Hi*-b reached the meninges by direct passage along the *olfactory* nerve, the pathway for the sense of smell, whose fibres originate in the brain and ramify into the upper part of the nasal cavity

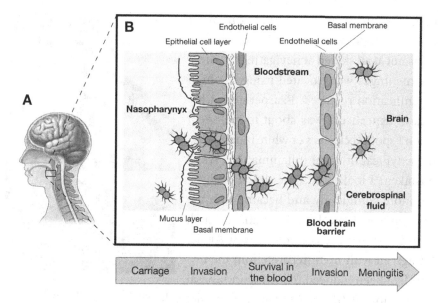

Figure 6.5   A schematic to show the multiple steps in which bacteria invade the linings of the of the upper respiratory airway (epithelial cell layer), enter the blood stream by traversing endothelial cells and disseminate around the body. The crucial invasion by bacteria of the barrier between the brain, meninges and cerebrospinal fluid (blood brain barrier) is not well understood. The importance of this pathway for causing meningitis is that antibodies in the blood can rapidly and efficiently eliminate bacteria thus preventing the occurrence of meningitis.

(Figure 6.4). Alternatively, most experts favoured a less direct pathway involving invasion of nasopharyngeal blood vessels or lymphatics. I was excited when I found that many of the rats had *Hi* bacteria in their blood (called *bacteraemia*) a few hours after inoculation. I did serial blood cultures and found that bacteraemia persisted for several days. This seemed promising; there was sustained infection of the blood following intranasal inoculation of *Hi-b*. But the million-dollar question was whether the rats had developed meningitis. Fortunately, my mentor Arnie Smith knew a skilled pathologist who worked nearby. To my intense frustration, I had to wait two months before I found a missed telephone call message from our collaborator's technician on my laboratory bench. The note said that I could pick up the stained brain sections from the Harvard Animal Research Centre.

Less than an hour later, seated anxiously at my microscope, I began looking through the couple of dozen slides, not at all confident that I would be able to interpret what I was looking at. At first, nothing seemed to make much sense as I was not at all skilled at navigating to where the meninges were located using the high magnification required. Exasperated at my lack of expertise, I was about to seek help when I spotted clusters of white blood cells,[f] those typical of acute inflammation. Was I mistaken? I looked at more slides, navigating more confidently and became sure the findings were typical of meningitis. Excited, I rushed down the corridor to tell Arnie. He agreed with my interpretation: meningitis had been induced following intranasal inoculation of *Hi*, an experimental model that offered numerous opportunities for further in-depth experiments. By examining the brain and meninges of animals at different

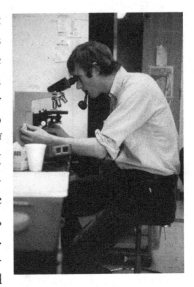

Figure 6.6 The author in his laboratory examining a sample of cerebrospinal fluid using microscopy ca. 1974.

times, it became clear from the location of the disease process that the bacteria had not tracked via the olfactory nerve but were reaching the brain via the blood stream and entering through a structure called the choroid plexus.[g,3]

I knew now that I had the motivation and passion for this kind of investigation. I was prepared to work long hours and to stick to a project even when most of the time there were setbacks and the experiments didn't work. I was completely hooked on research. A career that involved only clinical practice without a substantial commitment to research would not do. I wanted to be a clinician-scientist.

---

[f] Called neutrophils. See also Chapter 1.

[g] Finding bacteria in the blood alone did not allow distinction between the ascending nasopharyngeal or blood stream routes. If the bacteria invaded via the olfactory route to reach the meninges, bacteria would still enter the blood because cerebrospinal fluid drains directly into the blood stream. But the characteristics of the pathological findings showed that the bacteria reached the meninges via the blood stream not by the ascending route. Later work in primates consolidated these findings (see Ref. 3).

References

[1] Fothergill, R.D. and Wright, J. The relation of age incidence to the bactericidal power of blood against the causal organism. *Journal of Immunology*, 1933, 24:273–284.

[2] Medawar, P.B. *Advice to a Young Scientist*. Basic Books, 1979.

[3] Scheifele, D.W., Daum, R.S., Syriopoulou, V. *et al*. Haemophilus influenzae bacteraemia and meningitis in infant primates. *The Journal of Laboratory and Clinical Medicine*, 1980, 95:450–462.

Early Research on *Hi*-b Vaccines

My research on an animal model of *H. influenzae (Hi)* meningitis was a modest part of the larger and more ambitious research agenda to develop a vaccine to prevent *Hi* type b *(Hi-b)* meningitis. But the Boston Children's Hospital scientists were not alone in this quest. Two scientists based at the National Institutes of Health in Bethesda, Maryland — John Robbins and Rachel Schneerson — were also attempting to develop a vaccine, although it was not until the rival groups attended an Annual Paediatric Research meeting in 1972 that each became aware of the other's work. John Robbins, the son of Jewish immigrants from Brooklyn and a graduate of New York University, had experienced at first hand the impact of *Hi*-b during his time as a junior doctor working in Florida. Highly intelligent, streetwise and charismatic, John had extraordinary scientific vision. He came across as a domineering, politically incorrect workaholic but someone who showed huge generosity to those around him. His colleague Rachel Schneerson, a Polish-born Jewish immigrant and distant relative of Grand Rebbe Menachem Schneerson, the most prominent Lubavitch rabbi in New York City, complemented Robbins with her hard-nosed, technical brilliance in the laboratory; she was fiercely loyal and dedicated to Robbins's research.

Both the Robbins–Schneerson and Smith–Anderson teams had approached the problem in a similar way following in the footsteps of the research on pneumococcal and meningococcal polysaccharide vaccines (see Chapter 5). The first step was extraction and purification of the *Hi*-b capsular polysaccharide, polyribose-ribosyl phosphate or PRP (see Chapter 6 (Figure 6.2), p. 58). The second was to develop an accurate assay to measure blood levels of antibodies to the polysaccharide in laboratory animals and humans. Both rival groups had achieved these aims independently by 1972.

Now came the key question. Could PRP, used as a vaccine, induce antibodies and protect against disease? To investigate this idea, adults, children and infants were immunised with different amounts of PRP. The results were clear-cut. Strong antibody responses occurred only in children older than about 18 months. Younger babies didn't respond or if they did, the level of antibodies was not enough to provide protection.[a,1] This was a hammer blow because about three quarters of all cases of *Hi*-b meningitis occur before the age of two years. Various efforts were made to improve matters — such as immunising

Figure 7.1   Rachel Schneerson (left) and John Robbins (right).

with larger molecular sizes or amounts of PRP. There were no improvements. It seemed that the *Hi*-b meningitis vaccine project had reached an *impasse*.

It coincided with my being seconded to Lima, Peru, to gain further clinical experience in managing infections in a setting where many of the children were severely malnourished. But I had an idea that I thought was worth trying. I thought it might be possible to boost *Hi*-b antibody responses in very young children by giving PRP at the same time as the diphtheria, *pertussis*, tetanus (DPT) vaccine that was routinely given to all infants. I had read papers showing that the pertussis (whooping cough) component of the vaccine had the potential to enhance antibody responses. My plan was to compare infants given a single injection of PRP mixed with DPT compared to PRP and DPT given as separate injections.

I vividly recall my arrival in Lima. My senses were overwhelmed as sirens wailed, horns blared; buses, cars, scooters and mopeds revved their engines. My taxi ascended to Miraflores, a wealthy suburb of Lima, where I had been booked into lodgings. I recall later that day taking a relaxing walk past the elegant houses, often gated and guarded, and through a beautiful park with

---

[a] One of the uses of the animal model was to estimate the protective level of serum antibodies needed to protect against meningitis. The value obtained was 0.15 µg/ml blood. These data were important in the later submission to the FDA that led to licensure of the first *Hi*-b conjugate vaccine (see Ref. 1).

an ancient olive tree with a plaque to indicate that it was planted at the time of Pizarro, the fifteenth-century conquistador.

The Research Institute in Lima was conducting internationally acclaimed research on malnutrition and the Director, George Graham, had given his blessing to my project. However, I soon discovered that my carefully conceived research plans had been somewhat derailed by his daughter Marianne, a young woman who had recently graduated from the University of Pennsylvania and returned to her former home to work as a volunteer for a year. She had been assigned to help carry out a mass immunisation campaign to provide DPT vaccine to poor children in the *barriadas* of Lima. Her project had been so hugely successful that virtually all infants in the catchment area had been vaccinated, so there were hardly any unimmunised children who could be recruited into my study. I was impressed by Marianne's organisational skill as well as the natural, confident *rapport* with which she interacted with staff and patients, chatting in fluent Spanish, her long dark hair caught in a pigtail that fell over one shoulder. I was instantly smitten. I asked her if we could meet, on the thinly contrived pretext of discussing my research study, and we went to a popular pizzeria in the centre of Lima. Our first meeting was not the success that I hoped for. Marianne related to me her negative impressions of the several people who had come from the US to do research as well as her reservations about my own project. She was sensitive to the possibility that impoverished Peruvians, especially young children, were being exploited for research that was aimed more at advancing the careers of medical scientists than improving the health of the indigenous population. It seemed that she had no interest in my research or me; my pride was hurt and my plan to involve her in my project appeared to be a non-starter.

A few days later, however, matters changed. A school friend of Marianne's visiting from the US contracted a severe case of hepatitis and was admitted to a public hospital. The young woman was very sick, and her care was not going well. Distraught, Marianne asked me to help. I was able to arrange for her friend's transfer to the Anglo-American Hospital, where she was properly treated. Marianne now agreed to help with my research and over the next several weeks we worked together to recruit enough unimmunised children (who would benefit from being given the DPT vaccine). But when the assays were carried out many weeks later, the poor immune responses of infants to PRP were not improved by combining DPT with PRP.

Figure 7.2   The author with some of the mothers and young children participating in the PRP vaccine study conducted at the Nutrition Research Institute, Lima, Peru, 1973.

It is worth noting that our recruitment process in 1973 was unsophisticated compared to what would be required today. Marianne obtained the Institute's records of children who were part of an ongoing nutrition research programme to identify those that had not been immunised with DPT. We drove around the *barriadas* in a battered VW beetle to locate the children (no simple task in many cases) and ask permission from the parents for their child to be included in the study. I had "ethical permission" from an informal meeting with the Institute's Director of Clinical Studies based on a brief, hand-written protocol that I had submitted the day before. Consent from participating families on behalf of the enrolled child was verbal. The child came to the clinic where I obtained a blood sample, gave the immunisation and then arranged a follow up visit to get a further blood sample. It was my first experience in "field work" research and gave me huge satisfaction. It is worth noting just how hugely research procedures have changed since 1973. It is utterly inconceivable that such a study could be undertaken in this fashion today.

Meantime, the relationship between Marianne and I had deepened, but at the end of February 1973, a few days before I was due to leave Lima to return to Boston, she received an offer from the Peace Corps to volunteer in Paraguay for three years. We realised that we were unlikely to maintain our close relationship if she took up the offer. In what was for both of us a precipitate decision, we resolved to get engaged and to live together in Boston.[b]

After returning to Boston from Peru in early 1973, the mood in the laboratory over the PRP vaccine was still gloomy. Further results from a large trial (50,000 children) in Finland had only provided further evidence of the refractory antibody responses of infants to PRP.[c] To add to this, a smaller trial of Gotschlich's meningococcal polysaccharide vaccines in infants showed that antibody results were also markedly reduced in young infants.

The problem of poor immune response of infants to PRP was given a high priority for discussion at a scientific workshop[2] at the National Institutes of Health in Bethesda in 1973. A key contribution, strongly endorsed by the meeting's convener William Paul, came from Elvin Kabat, who almost did not attend the meeting because he was anxious about his allergy to tobacco smoke. In the 1970s, cigarette smoking at meetings was still common even among biomedical scientists. Kabat highlighted the research at the Rockefeller Institute that had been done many decades previously when Avery and Heidelberger were stymied by their failure to get efficient antibody responses to pneumococcal polysaccharides.[d] It might seem that due diligence in reading the extensive literature would have made the idea of coupling PRP to a protein an obvious way to solve the problem. But these scientific papers published in the 1920s and 30s had been largely forgotten. It's a well-known problem among scientists that the older literature is often neglected and even thought to be "old hat." It was fortunate that Elvin Kabat, a former graduate student of Heidelberger, was so familiar with the research that Avery and Goebel had done in the late 1920s. He suggested that their methods might enhance the immune responses to Hi-b in infants. Inspired as it would prove to be, there was at the time no

---

[b] Marianne's parents lived in Baltimore, Maryland, and we were married there in 1973, little knowing that the following year I would become a faculty member at Johns Hopkins Hospital and the city would be our home for ten years.

[c] There had still been some faint hopes that the weak antibody responses in infants (based on laboratory tests) might still be enough to protect in clinical practice especially because in children older than 18 months, the PRP vaccine (made in Porter Anderson's laboratory) was highly protective and this would become very important as discussed in Chapter 13).

[d] See Chapter 5.

evidence that coupling the polysaccharide to protein would overcome the refractory immune responses of infants. Only that the conjugates induced antibodies to polysaccharides more efficiently in adult laboratory animals. Nonetheless, it was the first time that I recall specific mention of an idea that resonated especially strongly with John Robbins.

After the workshop, Robbins read and re-read the studies of the Rockefeller scientists. Of course, in the ensuing 50 years there had been enormous progress in understanding the biology of antibodies. It had been discovered that their efficient production required interactions between two distinct kinds of immune cells, called B and T *lymphocytes*. Pure polysaccharides do not interact with these lymphocytes very efficiently, especially in the very young whose immune system is immature. So, although these thwarted antibody responses can just about pass muster in older children and adults, they are wholly inadequate in infants. The text of one of the key Rockefeller papers was etched in Robbins's mind.

> "Simple sugars, which by themselves are not antigenic will, when coupled to a protein, stimulate the formation of antibodies that are specific for the sugar used ... changing completely the antigenic specificity of the respective compounds."

By chemically linking PRP to a protein, Robbins reasoned that perhaps he could improve the refractory antibody responses in infants. He and Rachel Schneerson set about developing the required chemistry, a task that was much more complicated than had been the case for the Rockefeller scientists in the 1920s. The aim then had been to obtain antibodies to pneumococcal polysaccharides for laboratory research. But in the 1970s, Robbins was trying to develop a vaccine that would have to be completely safe when given to healthy infants and young children. As the head of a department (Bureau of Biologics) within the National Institutes of Health that was responsible for the safety of medicines, he knew intimately the stringent oversight required by the regulatory authorities. Further, the required conjugation chemistry had to be highly efficient so it could be easily and reproducibly scaled up by vaccine manufacturers. Many thousands of doses of the vaccine would be needed to carry out the necessary human trials. After this, millions of doses would be needed if the vaccine was to be widely implemented. It was going to need years of research and substantial funding.

Realising the enormity of the challenge, Robbins decided to share his plans with David Smith and Porter Anderson in Boston and encourage them to pursue the Hi-b conjugate idea. Despite their rivalry, Robbins sent them a summary of the relevant literature on conjugates and proposed that it would make sense for both groups to share their results. In particular, he suggested communicating the outcome of experiments that hadn't been successful so that neither group would waste time on unprofitable avenues of research. Scientists don't always act in this altruistic fashion but, throughout his career, this type of *largesse* was characteristic of Robbins. Porter Anderson was happy with Robbins's proposal, but the highly competitive David Smith was unwilling to reciprocate.

At the time, David Smith was a deeply troubled person. His bid for tenure at Harvard University had been rejected and his family life was turned upside down. Literally overnight, his wife had suddenly left the family home and David had become a single parent of three young girls. Angry and alienated, this deeply unhappy period in his life resulted in him making the decision to leave Boston. He had been head-hunted to become the Chairman of Paediatrics at the prestigious Rochester University Medical School in New York State, a post with huge administrative responsibilities. Porter Anderson moved with him and by mutual agreement, took charge of the laboratory research on Hi-b.

References

[1] Anderson, P.W. A lucky career in bacterial vaccines. *Human Vaccines and Immunotherapeutics,* 2012, 8, 4:420–422.

[2] Paul, W.E. New Approaches for Inducing Natural Immunity to Pyogenic Organisms. Department of Health, Education and Welfare Publication No. (National Institutes of Health) 74-553. U.S. Government Printing Office, Washington, DC, 1973, 157–166.

Chapter

8

Johns Hopkins Hospital

My fellowship at an end, I left Boston in the summer of 1974 to take up a staff position at Johns Hopkins. After marrying in the autumn of 1973, Marianne and I bought a semi-detached red-brick in the neighbourhood called Guilford, close to the main Johns Hopkins University Homewood campus. Marianne began her graduate training to become a teacher and I took up my post as an assistant professor at Johns Hopkins Hospital, a position in which I was hired to provide expertise in infectious diseases, my major clinical responsibilities being at the affiliated Baltimore City Hospital.[a] The Johns Hopkins Medical Institutions are in the inner city where, in the nineteenth century, newly arrived African Americans made their homes and worked in the city's shipyards. The later generations populated the inner-city neighbourhoods and, by the 1970s, faced massive unemployment following the economic decline of the docklands and the demise of the Bethlehem Steel Company. The ensuing political and police corruption, violence and drug-dealing were popularised in the HBO cable TV series *The Wire*. Located in the eastern part of the inner city, the Johns Hopkins Hospital with its lofty dome is a striking landmark, seemingly expressing the aspirations of its thousands of physicians and other health workers. Beneath the dome is a huge statue of Christ fashioned after Thorvaldsen's original work in Copenhagen.[b,1] One could not help feeling inspired and privileged

---

[a] The Chief of Paediatrics was Harold Harrison, a superb clinician who made major contributions to childhood diseases involving calcium and vitamin D. His sons, Stephen and Rick Harrison, became lifelong friends. Rick's daughter, Melissa, would many years later spend a year in my laboratory in Oxford. These friendships through academia and their consequences are one of the joys of science.

[b] A marble statue completed in 1838 in the Lutheran Church in Denmark's A Church of Our Lady in Copenhagen.

(a)   (b)

Figure 8.1   (a) The Johns Hopkins Hospital in 1905. Credit: Wellcome Collection (CC BY 4.0). (b) Bertel Thorvaldsen's Christus — under the Dome at Johns Hopkins Hospital. Credit: Art Anderson (CC BY-SA 3.0).

to be part of an institution that since the early part of the century had been recognised as the model for all North American medical schools.[2]

After eight years of post-graduate training in the UK and US after qualifying in 1966, it was exciting to have my first faculty appointment. In addition to my specialist work caring for sick children suffering from infections, I was often on call for all new general paediatric admissions. While doing ward rounds and out-patient clinics, I was also teaching medical students and trainee paediatricians. Despite this, I was determined to make time to continue my research.

From the outset, I made a point of establishing strong links with John Robbins and his large research group at the National Institutes of Health (NIH) in nearby Bethesda, just outside Washington, only an hour away.[c] John showed great interest in my research; his influence and encouragement were to prove critical in my further development as a research scientist. I was anxious that the experimental animal model of meningitis that I had developed in Boston might not be reproducible in this new setting, including the different animal facilities in Baltimore. The only laboratory space available to

---

[c] Although Johns Hopkins was a thriving research community, nobody was doing research specifically on bacterial meningitis. However, I received a lot of encouragement from three neuroscientists: Guy McKhann, Neil Nathanson and Richard T. Johnson. All were internationally known for their research on viruses and the brain.

me was in the basement of Baltimore City Hospital, a run-down, cockroach-infested facility that lacked many items of essential equipment. I had been given limited research funds on my appointment, enough to purchase some basic equipment and to have some animals freighted to my laboratory. In Boston, I had used classic pathology on brain sections to document meningitis; it was time-consuming and costly. It was therefore essential to develop a technique to obtain cerebrospinal fluid (CSF) from the very small animals, the same method used to diagnose meningitis[d] in the clinic. As a paediatrician, I had done lumbar punctures routinely on small babies. But my laboratory rats weighed only about 12 grams, hundreds of times smaller than human infants. To develop a method for obtaining CSF, I needed access to a dissecting microscope and that meant getting help from a colleague with a fully equipped laboratory at Johns Hopkins, a couple of miles from the City Hospital. I initiated the infection of animals in the City Hospital laboratory and then transferred them to Johns Hopkins where I could work on the technique to obtain CSF. This meant transporting the animals by taxi. When my cab driver discovered that I had a litter of infant rats, he was irate and ordered me to get out of the taxi. I found myself abandoned on the streets of inner-city Baltimore on a humid, hot day in possession of a cage of rodents. There are times in research when one questions one's sanity. As they say in the US: go figure.

It took several months, but eventually I perfected the method for obtaining CSF from my animals.[e,3] I could now make correlations between the numbers of Hi-b bacteria in the blood and the occurrence of meningitis. I was excited by the results and called John Robbins. To my delight, he was enthusiastic and invited me to give a departmental seminar to his research group. I had found that meningitis only occurred when the numbers of bacteria in the blood exceeded a critical density. This threshold was 1,000 or more bacteria per millilitre[f] of blood and suggested a basis for the propensity of Hi-b to cause meningitis. The idea

---

[d] Lumbar puncture was first introduced into clinical practice in 1891 by the German physician, Henry Quincke.

[e] The technique used to obtain cerebrospinal fluid was to insert a small disposable needle into the *cisterna magna*. This yielded from 10–25 µl of fluid, sufficient to carry out microscopy, culture and counts of white cells and bacteria (see Ref. 3).

[f] Corresponds to about one-fifth of a teaspoonful.

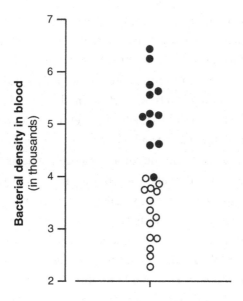

Figure 8.2 Each circle represents the bacterial count in the blood of an animal 48 hours after being infected with *Hi*-b bacteria. A cut off threshold of around 10,000 bacteria per ml of blood distinguished those animals that were found to have meningitis (●) as compared to those animals in which meningitis did not occur (O). The vertical axis shows bacterial density (thousands of bacteria per ml of blood) estimated by taking blood cultures that were plated on to solidified growth medium in *petri dishes*. After overnight incubation, the numbers of bacterial colonies could be counted to estimate the density of bacteria in the blood.

was that the type b capsule conferred a heightened capacity to cause high levels of bacteria in the blood stream, a property denied to other capsular types of *Hi*.[g]

My seminar at the NIH was well received, but to pursue an academic career in the US, I had to get research funding. Indeed, I had every reason to be anxious about my future. I had now been in the US for more than four years and returning to the UK as a hospital-based paediatrician was no longer tenable. My career pathway in the US had diverged significantly from the established trajectory for clinical academics in the UK. Besides, Marianne was thoroughly enjoying graduate school where she was training as

---

[g] As I mentioned earlier, it was Margaret Pittman who had first identified the six different capsular polysaccharide structures of *Hi* and had noted how almost all cases of meningitis were caused by type b stains. Now aged 74 and retired, she attended my seminar after which she and John Robbins took me out to lunch, a huge thrill that I still vividly recall.

an elementary school teacher. If I was to progress up the ladder at Hopkins, I had to make my mark through excellence in research. There is a harsh but true adage: publish or perish; but to do the rigorous research required for publication in prestigious scientific journals, I had to have adequate research funding.

I thought I now had an exciting hypothesis as to why type b, but not the other capsular types of *Hi* bacteria, caused meningitis that could be further investigated in the animal model. Robbins agreed, adding, "of course, I have my own ideas and will soon be publishing a paper on it." Typical Robbins; always wanting to be a step ahead, fiercely competitive but also generous and friendly. "Tell you what though, I'll give you some *Hi* strains from our collection that make different types of capsule. I am happy for you to use them for your studies."

Over the next months, I prepared what I thought would be a competitive grant application to submit to the NIH and asked John Robbins to look at it. I wasn't prepared for what happened next: "Rich, your application won't be funded," he said authoritatively. I could have wept. After the agonies of weeks of writing a research proposal, this was a devastating blow. I felt utterly miserable as I waited for what seemed like an eternity for John to finish his discussions with other research scientists so that I could sit down with him, one on one, to discuss my proposal. "Well, actually, the science is fine," he said as his face broke into the typical Robbins's mischievous look, "But you have to change the title! As it is now, it will go to the NIH section on bacterial infections and their current budget is already overspent. Your title must start with something that emphasises the *brain*. How about *Central Nervous System Induced Injury by Hi-b*," he suggested. "With that title, it will go to the neurology section and they have plenty of money." I cannot describe the enormity of my relief. From fearing the worst, I was now light-headed at the unexpected endorsement. I took John's advice, a good example of what is called *grantsmanship*[h] and went ahead with submitting the application. It would be several months before I learned the outcome.

Meantime at Johns Hopkins, I was about to face a terrifying ordeal: my inaugural presentation at what is called Grand Rounds, a weekly meeting

---

[h] The "art" of acquiring financial grants through the process of grant writing. The term is typically used when referring to the skills necessary to secure peer-reviewed research funding, but it can also apply more broadly to the overall field of fundraising from private foundations, community foundations, governments, and other grant-makers.

involving the entire faculty, trainees, students and many visitors. A recent case of meningitis was presented, following which my task was to discuss and analyse the child's illness and talk about new developments in the management and prevention of *Hi*-b meningitis. I was literally shaking as I went to the podium to begin my presentation trying hard not to be distracted by the sea of faces. A central theme of my discussion was how the *Hi*-b capsular polysaccharide was both friend and foe. *Friend* because of its potential as a vaccine, *foe* because the capsule is the major virulence factor responsible for making the bacterium a deadly pathogen. At question time, a senior professor (Barton Childs) challenged me:[i,4] "... if a vaccine is developed using the *Hi*-b polysaccharide, aren't you worried that a change in the bacterium's outer coat could occur and then the vaccine won't work?" It was a profound question and I must explain why.

One of the most important ways through which vaccines mediate protection is by inducing antibodies. These are proteins made by special cells in our body called lymphocytes. A typical adult has around 50,000 different antibodies, each able to recognise and bind to one specific component of the many pathogens that we have encountered. So, each of us has a repertoire of antibodies that depends on our prior exposure to microbes.[j] Antibodies are highly specific in their recognition of structures on the bacterial surface. If this target changes, for example the capsular polysaccharide, then the specific antibodies do not bind and the bacterium escapes. Variation in microbial surface components (called *antigens*) is one of the commonest reasons why vaccines fail. It's an enormous and complex problem whose shadowy presence

---

[i] Barton Childs. 1916 2010. A paediatrician who defined the field of genetic medicine. His book is a visionary conceptualisation of the interactions of genetic and societal factors in human disease.

[j] This is called *adaptive immunity*. It requires exposure to the pathogen and takes time (several days or more) to develop. Thereafter, our cells remember the encounter (*immunological memory*).

So, what happens when we encounter microbes for the first time before there has been time for adaptive immunity to be in place? To solve this problem, our body has non-specific mechanisms (a first line of defence) that can be activated immediately to ward off microbes that the body has never encountered before. This so-called innate immunity triggers inflammation and other non-specific resistance mechanisms. A crude analogy is to imagine what happens when a building catches fire. It sets off the sprinkler system (innate immunity) that may keep the fire at bay before the alert goes out to bring in the specialist fire brigade (adaptive immunity).

can be considered "the elephant in the room" of *vaccinology*. Called *antigenic variation*, this diversity of the surface molecules of microbes will be at the forefront of discussions in later chapters.

I answered the senior professor by conceding his point and adding that since we didn't understand why *Hi*-b organisms are so much more deadly, it was indeed possible that other highly virulent capsular variants could emerge against which the vaccine wouldn't work. I was thinking that the question was a powerful argument in favour of the research that I had proposed to the NIH. Indeed, in the autumn of 1975, I learned that my research application had been approved. It was a huge relief. Even better, instead of the three years of funding that I had requested, I had been awarded almost $500,000 of research funding over a suggested time frame of five years — subject to satisfactory progress by the third year. It seemed scarcely believable; for the immediate future, my career was secure. The award changed my profile in the eyes of my senior colleagues at Hopkins and I was allocated generous laboratory space in the main research area of the Paediatrics Department. I shed no tears over leaving the cockroach-infested basement laboratory for the far superior facilities at Johns Hopkins.

References

[1] McCall, N. The statue of then Christus Consolator at the Johns Hopkins Hospital: Its acquisition and historic origins. *Johns Hopkins Medical Journal,* 1962, 151:11–19.

[2] Starr, P. *The Social Transformation of American Medicine.* New York Basic Books Inc. 1982.

[3] Moxon, E.R. Experimental studies of infection in a rat Model. Haemophilus influenzae: Epidemiology, immunology and prevention of disease. Sarah, H. Sell and Peter, F. Wright (eds.). Elsevier Biomedical, 1982.

[4] Childs, B. *Genetic Medicine: A Logic of Disease.* Johns Hopkins University Press, 1999.

Snake Venom, Bottlenecks and Genetics: The Challenges of Research

As a research fellow in Boston, having excellent research facilities had been something I took for granted. The past year had been a rude awakening; I had struggled because of the inadequate facilities and equipment in my make-shift laboratory at Baltimore City Hospital. Now, thanks to funding from the National Institutes of Health, I had been allocated a generously sized laboratory in the Johns Hopkins Department of Paediatrics. My working days were divided between looking after patients, teaching (medical students and junior doctors) and research. It was a huge privilege to be in a large and prestigious paediatric department consisting of so many brilliant and inspiring colleagues. Their passion and love of medical science left me in no doubt that combining the challenges of work in the hospital wards and clinics with laboratory research was enormously fulfilling. But being a clinician-scientist, a doctor who combines medical practice with laboratory research is a tough challenge. As has been succinctly articulated in a very honest article by one of my contemporaries, "... one quickly finds that basic scientists are sceptical about your scientific knowledge and ability while your clinical colleagues may not regard you as a top-notch clinician, with both groups viewing you with not a little suspicion."[1]

As in all walks of life, a little bit of good luck comes in handy. Across the corridor was the laboratory of another clinician-scientist of my own age, Jerry Winkelstein. Over the next several years, we shared the ups and downs of our research and collaborated on a range of projects. We also joined forces to set up a clinic for children with immune deficiencies where our respective specialty expertise in infectious pathogens and host immune mechanisms proved to be synergistic. We were an unlikely pair. I am a British version of what North Americans call a WASP (white, Anglo-Saxon protestant)[a,2] whereas Jerry's grandparents were Jewish immigrants from Lithuania who had settled in Syracuse, New York. Small in stature (whereas I am well over 6 feet in height), with wiry, black hair and a studious demeanour, Jerry disguised his natural anxiety about almost everything in life with an articulate, good-natured repartee and a penetrating intellect.

My good luck was that he was an expert on a set of proteins, called the *complement pathway*, critical to the immune clearance of bacteria. I have already described the key role of antibodies in eliminating pathogens by binding to components on the bacterial cell surface. But getting rid of pathogens requires a lot more than antibodies. As their name suggests, the complement proteins enhance bacterial killing. When antibodies bind to the bacterial surface, they activate a cascade of nine complement proteins in a domino effect. One after another, each component is summoned into action resulting in a lethal assault that kills the bacterium by two mechanisms. One involves the engulfment of the bacterium by white blood cells; the other weakens the bacterial membrane so that it sustains a lethal leak and the microbial cell dies by *lysis*.

To investigate its important role in meningitis, Jerry suggested that we deplete the complement proteins in the rat to see what effect it had. When I asked him how it was possible to selectively eliminate the complement pathway, he casually mentioned that it involved using cobra venom! To obtain the required non-toxic protein (called Cobra Venom Factor[b]), commercially purchased venom had to be purified through a series of chemical procedures. This required several days of dangerous work in the "cold room" adjacent

---

[a] The term was popularised by sociologist and University of Pennsylvania professor E. Digby Baltzell, himself a WASP, in his 1964 book (see Ref. 2). WASPs were the first immigrant group to settle in America in the late seventeenth and early eighteenth centuries coming from northern Europe, especially from Britain, Ireland, Germany and Scandinavia. One of my direct ancestors, George Moxon, emigrated to New England in 1637 and was the founder pastor of Old First Church in Springfield, Massachusetts.

[b] Cobra Venom Factor (CVF) is a non-toxic protein purified from cobra venom. It continuously and artificially drives the complement cascade to the point of depletion, a bit like leaving a torch on so that the battery runs down.

to our labs. Hazardous work indeed, and for a few days the corridor was littered with warning signs so that no unauthorised persons went into the "snake–pit" — as the cold room was temporarily renamed.

I have described in Chapter 8 (see Figure 8.2) the exquisite relationship between the density of bacteria in the blood and the occurrence of meningitis. When we depleted the infected rats of their complement components using the cobra venom factor, there was a substantial increase in the number of animals with meningitis. In the absence of the complement proteins, clearance of bacteria from the blood was impaired. As a result, more animals had numbers of bacteria in their blood that exceeded the critical threshold (a thousand bacteria per ml), so more cases of meningitis occurred, which was pleasing evidence in support of my hypothesis that the magnitude of blood-borne bacteria was a key determinant underpinning its occurrence. This was the first of several experiments that Jerry and I did together, a productive and formative time for both of us that remains one of the abiding memories of my time at Hopkins. When neither of us were on call for the clinical service, we would meet first thing in the morning in the cafeteria for coffee and doughnuts. It was through these meetings that so many of the ideas for my experiments were conceived and refined before being tested in the laboratory.

Outside of work, Marianne had graduated from Goucher College and was about to begin an assignment as a trainee teacher. We felt ready to start a family and were thrilled when testing showed that Marianne was pregnant. Our excitement was short-lived. The positive pregnancy test was in fact indicative of an aggressive cancer, a choriocarcinoma of the uterus, whose malignant cells release the same hormone as in pregnancy.[c] As a medical student in the 1960s, I'd learned about this cancer. Etched in my mind was the fact that it spread rapidly to the lungs and was usually fatal in a matter of months. Fortunately, in the ensuing years since I had been a medical student,

---

[c] Choriocarcinoma arises from a failed reproductive event in which only the father's DNA, from the sperm, enters an egg that for unknown reasons does not have any female-derived DNA. The male DNA confers the characteristic of very rapid growth on the faulty egg that, in the absence of DNA from the mother, is not held in check, resulting in a potentially cancerous state called a mole — because of the way it burrows into the uterine tissues. Fortunately, because the cells driven by the paternal DNA divide rapidly, they are particularly susceptible to drugs that interfere with cell replication and more than 90% of these malignancies can be completely cured with the drug *methotrexate*.

it had been discovered that most patients could be completely cured with the drug methotrexate. Of course, there were uncertainties; when the statistics indicate better than 90% success, the threat of being the one failure in ten casts a shadow of uncertainty that can only be dispelled by time. It took 18 months of treatment before Marianne was thought to be in the clear and a further year before we could try once again to have a child. Inevitably, throughout these desperately worrying months, my research took a back seat, but once Marianne was in remission, it got back on track in an exciting but bizarre fashion.

I was giving a lecture to a small gathering of infectious diseases researchers when I saw one of my colleagues nodding off in the front row of the seminar room. He had been up late preparing lectures for the medical students and slept soundly for a full 40 minutes of my talk. Just as I was winding up and offering to answer questions, my friend awoke and, still in a state of sleepy confusion, asked a question that left everyone, including me, completely baffled as to its relevance. Still subliminally engaged in thinking about his lectures to the medical students, a scientific article that he had read must have surfaced as he recovered from his mid-lecture nap. It reported an intriguing experiment in which mice were given tens of thousands of virulent Salmonella bacteria via the mouth. After the organisms entered the gut, they invaded the intestines and entered the blood stream. But analysis of the bacteria cultured from the blood showed that all of them were derived from one surviving bacterium. At some stage all but one of the thousands of bacteria that had reached the gut were eliminated leaving one lone survivor to multiply and cause an overwhelming infection. Why did just one bacterium succeed in making it through the gauntlet of host defences? This same pattern of "The Winner Takes It All"[d] occurred time and time again in different experiments involving large numbers of mice; it was the rule rather than the exception. It seemed amazing and implausible, yet when I read the original paper,[3] I was mightily impressed. So, I repeated the experiment using Hi-b instead of the Salmonella bacteria that had been used in the original, published study. To my astonishment, the results showed that after instilling the nasal cavity of my baby rats with millions of Hi-b organisms, infection of the blood and meninges resulted from survival of a single bacterium from the infecting

---

[d] ABBA's hit song of 1980. Both the song and my experiment give me goose bumps of pleasure.

inoculum.[e,f] I should add that my "sleepy" colleague was an energetic and brilliant "Brit" from Liverpool. Since the "single organism experiment" had been his idea, it was only right that Patrick Murphy should be a co-author on the resulting article,[5] although I had done all the experiments. Patrick's laboratory and office were in the Basic Science Building of the Medical School where we got together to write the paper. Taking a break for a cup of tea, he introduced me to one of the stars of the Genetics Department who had for many years been doing research on the mechanisms of transformation of *Hi*. This was my first meeting with Hamilton ("Ham") Smith, the third influential Smith[g] in my research career.

Figure 9.1   Hamilton ("Ham") O. Smith (1931–). Credit: Jane Gitschier/Public Library of Science (CC by 2.5).

In the late 1960s, Ham had identified a hugely important enzyme found in *Hi* bacteria, called a *restriction endonuclease*. It was a protein that cut DNA in a very predictable way — only at sites where there was a specific sequence of nucleotides, the building blocks of DNA. It was the first of many such enzymes, each with unerring, distinct specificities for cutting DNA. Using

---

[e] Called a *single cell bottleneck*, it was a novel insight into the mechanisms leading to meningitis.

[f] The reader may well wonder by what means this conclusion was reached with certainty. In a typical experiment, 20 animals were each inoculated via the nose with a mixture of two distinct variants of *Hi*-b. Nasal swabs showed that the animals were indeed colonised with thousands of bacteria around half of which were resistant and the other half sensitive to the antibiotic *streptomycin*. Two days later, blood cultures from each animal were obtained. Given that each animal was colonised with a mixture of the two variants, one might expect that both variants of the bacteria would also be cultured from the blood. But this was not at all what was found. Most blood cultures from individual animals consisted of a pure culture of either the resistant or sensitive variant, not a mixture. The only plausible (statistically valid) explanation was that the entire population of bacteria in the blood was derived from a single founder *Hi*-b bacterium. At what stage did this population bottleneck occur? It would be 40 years before an explanation for this result would be discovered (see Ref. 4).

[g] In Boston, there had been David Smith, one of the pioneers of the *Hi*-b conjugate vaccine and Arnold Smith with whom I had worked in developing the experimental model of meningitis.

these enzymes, any fragment of DNA could be mapped, just as a stretch of road can be described by distinct bus stops (analogous to the enzyme cleavage sites) along the route. This was the crucial breakthrough that initiated the era of what became known as the *new genetics*[h,6] involving cloning and *recombinant DNA*.[i]

I told Ham Smith about my interest in understanding the role of the *Hi*-b polysaccharide in causing meningitis. "Sounds like you need to be using genetics to get any further on this," he suggested, "Why don't you isolate the genes for making the capsular polysaccharide since these appear to be crucial to causing meningitis?" I cannot adequately describe the impact of this piece of advice. I tried hard not to show how taken aback I was, but it must have been obvious that such an idea was completely alien to the kind of research I had been doing. The truth was that I hadn't the faintest idea how to go about doing bacterial genetics. My training and laboratory skills at that time were completely inadequate. It left me feeling, not for the first time, excited but incompetent.

Over the following days I realised that here was a golden opportunity to move my research into a different gear, one that seemed so much more sophisticated than the classical animal experiments that I had been doing. Fortunately, in December 1977, I learned that my application for a National Institutes of Health Research Career Development Award[j] had been successful. It provided a guarantee of salary and research expenses for a further five years, the perfect opportunity to undertake the ambitious plan of isolating the type b capsule genes — although so much depended on whether I could persuade Ham Smith to take me on in his lab.

---

[h] A term used by Botstein to describe the mapping of DNA using restriction enzymes in his seminal paper.

[i] See Chapter 10, Figure 10.3.

[j] One of the remits of an RCDA was to encourage medical scientists to take on bold and innovative research projects for which protection from clinical and teaching responsibilities was an important provision.

References

[1] Kennedy, P.G.E. Life as a clinician-scientist: Trying to bridge the perceived gap between medicine and science. *DNA and Cell Biology*, 2015, 34(6):383–390.

[2] Baltzell, E.D. *The Protestant Establishment: Aristocracy and Caste in America*. Yale University Press, Newhaven, Connecticut. USA, 1987.

[3] Meynell, G.G. The Applicability of the hypothesis of independent action to fatal infections in mice given *Salmonella typhimurium* by mouth. *Journal of General Microbiology*, 1957, 16:896–904.

[4] Ercoli, G., Fernandes, V.E., Chung Wen, Y. *et al*. Intracellular replication of *Streptococcus pneumoniae* inside splenic macrophages serves as a reservoir for septicaemia. *Nat. Micro*. 2018, 3:600–610.

[5] Moxon, E.R. and Murphy, P.A. *Haemophilus influenzae* bacteremia and meningitis resulting from survival of a single organism. *PNAS*, 1978, 75:1534–1536.

[6] Botstein, D., White, R.L., Skolnic, M. and Davis, R.W. Construction of a genetic linkage map in man using restriction fragment length polymorphisms. *American Journal of Human Genetics*, 1980, 32(3):314.

Chapter

10

A Needle in a Haystack: Searching for Virulent Bacterial Genes

My life had been a "roller coaster" during 1975–1977 as a result of Marianne's cancer. My foray into the basic science of bacterial genetics and the joys of fatherhood would make 1978 a special year. Our first child, Christopher, was born in March 1978. It was a scarcely believable and joyful turn around in our lives. On a beautiful early-spring day in April 1978, we drove to see the flowering cherry trees around the Washington Basin by the side of the Potomac. Despite the beauty of the surroundings and the heady emotions of parenthood, I was preoccupied. "I'm thinking about my research" I said to Marianne in a tone of voice she had come to recognise. My mind was in another world — a state that might metaphorically be called *Brain Fever*. "Just enjoy the magic of the water and the blossom," Marianne interrupted. Not taking the hint, I started to tell her why the research I was planning to do was so exciting and important to me. It was a blatant example of my addiction to science, a mentality that Marianne has had to cope with throughout our almost 50 years of being married. On this occasion, I think that she would have been more than justified in pushing me into the water, but with her customary patience she let me banter on, not without some scepticism as to what her paediatrician husband was now spending so much time doing; wasn't my career supposed to be dedicated to looking after sick children?

I was thinking about how I could identify the genes for making the type b polysaccharide. In my laboratory collection, there was a strain of *Hi* that had spontaneously lost the ability to make the capsule. I reasoned that it must have a mutation in one of the several genes required to make it. Using transformation (see Figure 10.1), I had shown that the mutation

DNA transformation

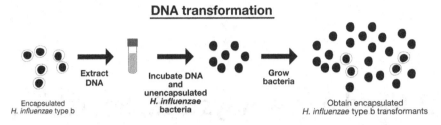

Encapsulated *H. influenzae* type b   **Extract DNA**   **Incubate DNA and unencapsulated *H. influenzae* bacteria**   **Grow bacteria**   Obtain encapsulated *H. influenzae* type b transformants

Figure 10.1   The process in which DNA extracted from one population of bacteria is taken up and incorporated into the genome of a different population of bacteria of the same species. In this example, the DNA from encapsulated *H. influenzae* type b bacteria (donors) is used to transform unencapsulated bacteria (recipients), resulting in a few bacterial colonies that express the type b capsule.

in the capsule-deficient variant could be corrected. This was a sound basis to move forward because molecular genetics could identify the relevant transforming DNA.

Ham Smith agreed to take me on in his laboratory where I was allocated a small amount of bench space in his crowded lab and my head of department granted me a period of two years during which I was relieved from carrying out teaching and clinical work. Although I was a relatively experienced clinician who had also been involved in years of medical research, I was now a wet-behind-the-ears apprentice learning the ropes from a cadre of young basic scientists whose research was curiosity-driven, not motivated by clinical observations. It was exhilarating but daunting; it had been more than a decade since I had done my degree in Natural Science at Cambridge, a period that had coincided with the 1962 Nobel Prize to two Cambridge scientists for the discovery of the DNA "double helix." But I had not kept up with the enormous progress in basic biological science and my lack of knowledge of genetics and DNA was embarrassing although, given my clinical background, I was aware that much of the progress had stemmed from research on antibiotic resistant bacteria. In the mid-1970s, there was mounting concern that pharmaceutical companies would not be able to discover new antimicrobials fast enough to keep pace with the rate at which bacteria were developing resistance to them. Because this increase in antimicrobial resistance was causing major problems in treating serious, often life-threatening infections, a great deal of research had been carried out on how bacteria become resistant. In many instances, the mechanism turned out to be through the transfer of small pieces of DNA that were quite distinct from the much larger, circular bacterial genome (Figure 10.2). Called *plasmids*, this ancillary DNA can be thought of as viruses that infect bacterial cells.

Bacterial viruses are a key part of recombinant DNA research. Using restriction enzymes — the first of which had been discovered by my new mentor Ham Smith, — the DNA of a bacterial virus can be "cut" into smaller, precise pieces. Under the right conditions, the viral DNA can be cut into just two fragments between which a "foreign" piece of DNA (from any other life form, plant insect, human, etc.) can be inserted. The three fragments are then joined using classical biochemistry. This cutting and pasting

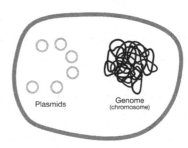

Figure 10.2 The bacterial genome (or chromosome) is usually a circular, double-stranded molecule ranging from 0.6 to 8 million nucleotides. In addition, bacterial viruses, including plasmids, may be present within the bacterial cell. These range in size from 1–200 thousand nucleotides, ten to 100 times smaller than the genome.

changes the information content or *genotype* of the virus because the host bacterium is now infected with "foreign" DNA that make proteins that alter the behaviour (*phenotype*) of the bacterium (Figure 10.3). The revolutionary

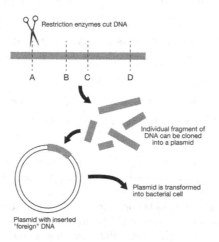

Figure 10.3 The basic steps in recombinant DNA technology involve the cutting of a piece of DNA into precise fragments. Different restriction enzymes (indicated by A, B, C and D) cleave the DNA through recognition of specific, small sequences of nucleotides. The separate fragments are inserted into a plasmid through "cutting and pasting" using various enzymes (including restriction enzymes) to create a recombinant plasmid containing an insert of foreign DNA. The genetic information contained in the foreign DNA confers novel functions to the recombinant plasmid.

changes in biology that resulted from the use of recombinant DNA technology raised serious ethical concerns. Science had changed from studying life as humans knew it to life as humans could make it. The potential of molecular genetics was so profound that, just two years before I joined Ham Smith's lab, there had been a moratorium on recombinant DNA research while the ethical implications of this powerful new technology were debated at a specially convened meeting — the iconic Asilomar Conference in 1975.[a]

Nonetheless, recombinant DNA brought about unprecedented scientific and commercial opportunities. Still in its earliest days, its utility for investigating the genetics of bacterial virulence offered an incredibly exciting opportunity. A seminar given by Stanley Falkow, a pioneer in the field of the molecular approach to investigating bacterial infections, provided my first opportunity to talk with someone who was actively pursuing my line of thinking. We shared ideas, a rather one-sided exchange since it was the brilliant Stan, not I, who was brimming with exciting insights. He explained how he and his colleagues had cloned a toxin from a bacterial pathogen that was responsible for severe diarrhoeal disease. I needed to move fast; it was obvious that he and other scientists interested in infections were on the same track.

Becoming proficient in this new discipline was not an easy transition for me. The "young bloods" in Ham's lab talked a different language, and for the first several months the presentations and lectures passed right over my head. But not for a moment did I regret my decision. I was determined to make headway into investigating why type b strains were so much more virulent than other capsular types, a question that could not be answered using the clinical isolates that I had been given by John Robbins.[b] The starting point

---

[a] The Asilomar Conference on Recombinant DNA was an influential conference, organised by the renowned scientist Paul Berg to discuss the potential biohazards and regulation of biotechnology, held in February 1975 at a conference centre at Asilomar State Beach. A group of about 140 professionals (primarily biologists, but also lawyers and physicians) participated in the conference to draw up voluntary guidelines to ensure the safety of recombinant DNA technology. The conference also placed scientific research more into the public domain, and can be seen as applying a version of the precautionary principle.

[b] In addition to differences in the genes for the different capsules (accounting for about 1% of the genome), there were also differences in the remaining DNA of the genomes of a, b, c, d, e and f capsular serotypes. To investigate the role of the b capsule genes in mediating the heightened virulence of type b strains, it was necessary to isolate each of the 6 distinct capsule loci and construct *isogenic* strains. This allowed experiments to compare the virulence of strains that differed only in the genes for capsule.

was extracting the genomic DNA from a *Hi*-b bacterial strain that had been cultured from the spinal fluid of a child with meningitis. Working with one of Ham's post-docs, I used restriction enzymes to cut up the two million nucleotides of the total *genomic* DNA of our chosen *Hi*-b bacterium into thousands of smaller fragments. Each of these fragments could be spliced into one of the specially modified bacterial viruses to make a *DNA library* (Figure 10.4). The idea was that at least one of the recombinant viruses making up the library would have an insert containing DNA required to restore the production of type b polysaccharide to my capsule-deficient mutant. The problem was how to find the "needle in the haystack."[c] Fortunately, there was a solution.

When grown on solid media, bacterial colonies of the mutant were a dull opaque grey in appearance, whereas colonies that had been restored by transformation to make the b capsule had a bright, shiny iridescence. Using a lamp at just the right angle, hundreds of colonies could be screened in a relatively short time. I was searching for the very rare iridescent colonies, nick-named "stars of Bethlehem" among the lacklustre capsule-deficient colonies.

The work was tedious; the library consisted of thousands of recombinant viruses, each of which had to be tested to see if it could transform the mutant. Plate after plate had to be inspected and it was easy to lose concentration. Then came the moment when, after many days of

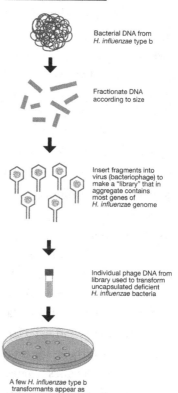

Making a "Gene" library

Bacterial DNA from *H. influenzae* type b

Fractionate DNA according to size

Insert fragments into virus (bacteriophage) to make a "library" that in aggregate contains most genes of *H. influenzae* genome

Individual phage DNA from library used to transform uncapsulated deficient *H. influenzae* bacteria

A few *H. influenzae* type b transformants appear as iridescent colonies when grown on agar plates

Figure 10.4   The several steps in cloning DNA form the genetic region for type b capsule production from *H. influenzae*. The whole genome of a type b strain was fragmented and the pieces of DNA inserted into bacterial viruses (called bacteriophages). DNA from these phages, each containing different pieces of the *Hi* genomic DNA, were used to test whether they could transform a capsule-deficient mutant *Hi* so that it regained capsule expression.

---

[c] The amount of DNA was estimated to be about 1% of the genome.

discarding the boring, negative plates studded with non-iridescent colonies, I saw a few "bright" colonies — stars of Bethlehem. I blinked, turned away, rested my eyes, and looked again. I felt a bit like the sailor in the crow's nest who, after days of tedious scanning of the flat, unvariegated horizon spies the unmistakable contours of *terra firma*. But, almost immediately, doubts took over. Would these shiny colonies prove to be rogues or princes? Perhaps the plates had been contaminated. My research assistant at the time, Carla Connelly, retrieved the tube of recombinant virus containing the fragment of DNA that had produced the positive result and we repeated the transformation. We also tested two other fragments from the library that had tested negative. I can still remember the agonisingly tense moments the next morning as I opened the incubator after the transformed recipient bacteria had been growing overnight. The positive result was confirmed. Relief and huge excitement — I was literally shaking, unable to stay calm. I rushed across the corridor to tell my scientist friend Jerry that I had cloned a piece of DNA that was involved in making the type b capsule. It was a truly exceptional day; there are only a few like this in your career as a scientist. In the late seventies, only a small number of bacterial virulence factors had been cloned and all were for proteins expressed on the bacterial cell surface. My finding was different. I had identified one or more genes for the enzymes needed to synthesise the type b polysaccharide. Now the longer-term goal was to isolate the complete region of DNA containing all the genes for making the capsule, a task I knew would take many years. But the first step had been accomplished.

In the meantime, I carried out an important experiment using my experimental animal model to show that transferring the cloned DNA into the attenuated *Hi* strain did indeed completely restore its ability to cause meningitis. I was now ready to discuss the successful results with Ham Smith who was thrilled but, in typical fashion, modestly declined to take any credit. In fact, he had not only suggested the idea but had provided me space and support in his laboratory. I had taken at least the first step towards answering the tricky question from the professor who had asked why *Hi*-b organisms are so deadly and whether highly virulent capsular variants might emerge against which a polysaccharide vaccine would be ineffective. Later work would show that transformants making the five other capsular polysaccharides did not cause meningitis. Type b strains had unique properties.

In October 1998, I awoke to some astounding news on the radio. Ham Smith had been awarded the Nobel Prize for Medicine or Physiology along with his Johns Hopkin's colleague, Dan Nathans, and the Swiss scientist Werner Arber, for their research on restriction enzymes. As I was involved that morning in clinical work, I could only make my way to the Basic Sciences Building to join in the celebrations later. When I arrived, I managed to squeeze myself into the packed lecture theatre where there was a press conference. A reporter was asking what *good* the discovery would do. Ham smiled and looked diffidently at Nathans, "this depends on some common understanding of what 'good' is," Nathans said. Ham added that "it'll be great to get to know how genes work — with understanding comes progress." I could not help smiling to myself. I had taken a first step towards a better understanding of the genetic basis of one disease: meningitis. More broadly, molecular genetics and recombinant DNA technology were changing the scientific basis of clinical practice. The "new genetics" was as applicable to pinpointing mutations responsible for human diseases such as breast cancer as it was for identifying the virulence genes of microbes.

Promotion: New Opportunities and Challenges

In 1979, aged 38, I was made Head of the Division of Paediatric Infectious Diseases at Johns Hopkins. For more than five years, I had had substantial protected time for my research, but this appointment brought about a major transition in my career with greater clinical, teaching and administrative responsibilities. I could no longer devote so much of my own time in the laboratory to progress the research on the work on the type b capsule genetics. The recombinant DNA library had identified just a small part of a larger genetic region (totalling about 1% of the complete genome). Delegating the project proved relatively simple as Susan Hoiseth,[a] who had just completed her PhD in bacterial genetics at Stanford University, had just joined my laboratory. Her flair was apparent immediately and very quickly she made an important discovery. The DNA that I had identified from the genomic library was part of a duplication, a high-frequency genetic switch that resulted in some of the *Hi*-b bacterial cells ceasing to make the type b capsular polysaccharide. It was puzzling and made no sense to me at the time although Susan thought (correctly as it turned out) that this was a mechanism through which *Hi*-b could improve its ability to colonise the nose and throat. The *Hi*-b genetics project was in excellent hands.

One of the most enjoyable facets of my new role was responsibility for a National Institutes of Health programme for training specialists in paediatric

---

[a] Susan would later become a lead scientist in the molecular biology team of Pfizer (Pearl River, New York, USA) This vaccine company would later develop safe and effective vaccines against meningococcal and pneumococcal meningitis.

infectious diseases. The fellowship programme stipulated that, in addition to clinical experience, each trainee had to complete at least a year of laboratory research. So, in addition to the time spent supervising their clinical activities, taking on these extremely bright and hard-working young paediatricians also meant a change in the *modus operandi* of my lab as I selected and supervised specific research projects for them. There were also many requests to take on overseas trainees and occasional graduate students in the Johns Hopkins MD/PhD programme (the so-called mud-fuds).

This provided a perfect opportunity to start some new research projects to investigate surface molecules, other than the type b capsular polysaccharide (PRP), that were implicated in causing meningitis. Although the role of PRP in promoting survival of *Hi*-b organisms in the blood stream was well established, the contribution of other bacterial surface components, for example in colonisation and invasion of the upper airways and meninges, was wide open. Using the library of recombinant viruses to search for the DNA involved in the PRP synthesis, I had found, but not yet investigated, a DNA insert that transformed *Hi* to produce colonies that had a striking white appearance. This suggested a change in one of the surface molecules of *Hi* other than the capsule. Here was something interesting and I assigned the project to a visiting Swiss doctor who had been seconded by his professor to my lab to gain research experience with which to complement his outstanding clinical skills and expand his understanding of infectious diseases.

Analysis of the mysterious transformant showed that it had an alteration in a surface macromolecule called *endotoxin*, a completely distinct molecule from the type b capsular polysaccharide (Figure 11.1).

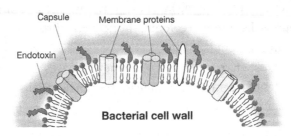

Figure 11.1   The major bacterial cell outer membrane components. Endotoxin, capsular polysaccharide (grey shading) and membrane proteins are depicted. These molecules are important in determining the virulence of the pathogens causing meningitis but are also the major targets for protective antibodies, the basis of safe and effective vaccines.

When the infant rats were infected with this transformed opaque *Hi*-b strain, none developed meningitis. We discussed the findings with John Robbins who was surprised (and sceptical) that an *Hi* bacterium expressing normal amounts of the type b capsular polysaccharide wasn't virulent. After months of careful studies, my visiting researcher (André Zwahlen), confirmed the importance of endotoxin in *Hi* virulence. These results suggested that the type b capsule was necessary, but not sufficient, to cause *Hi*-b meningitis. It was the start of many further years of research in understanding the subtleties of endotoxin in the pathogenesis of meningitis as well as its potential as a vaccine candidate.[b]

Also contributing to this research on endotoxin was a paediatrician (Lorry Rubin) on the NIH Fellowship programme who was spending substantial blocks of time on the clinical service, a good example of how valuable it can be to combine lab research with patient care. As specialists in infectious diseases, we were asked to help in the care of a baby girl with *Hi*-b meningitis who was not getting better despite antibiotic treatment. Initially, she had been treated with the antibiotics, *ampicillin* and *chloramphenicol*. After the hospital laboratory tests showed that ampicillin alone was adequate treatment, the chloramphenicol was stopped. But the infant became feverish and irritable and a repeat lumbar puncture showed persistence of meningitis. So, what had gone wrong?

It had been known for many years that some *Hi*-b bacteria can be resistant to ampicillin by making an enzyme that degrades the antibiotic. The gene for this enzyme resides on one kind of bacterial virus, called a plasmid,[c] that infects *Hi*-b bacteria. The standard test performed by the hospital microbiological laboratory to detect this resistance plasmid was negative. On investigation in my laboratory, Lorry found that the *Hi*-b strain from our patient had a different plasmid, one not previously described, so that the routine testing had not detected this novel enzyme that degraded ampicillin. Meantime, the baby had been treated successfully by re-starting chloramphenicol and she made a complete recovery. But here was a clear warning of a potential problem that could compromise successful treatment. The discovery of this new plasmid

---

[b] See Chapter 15.
[c] See Chapter 10, Figure 10.2.

and its implications for treating *Hi*-b meningitis were published in *The Lancet*, a real *coup* for my trainee.[1]

Another memorable example of how my trainees were changing my life came late one evening when my telephone rang and I heard the familiar voice of Mathu Santosham. Having completed his fellowship with me, he had joined the School of Public Health at Hopkins and had been posted to White River, a native American reservation in Arizona, to reduce the appalling mortality from diarrhoeal disease in indigenous American Apache natives. The reason for his phone call was to tell me that *Hi*-b meningitis was a major problem among Apache children. Indeed, further investigation showed that the rates of *Hi*-b meningitis were more than a hundred times higher than elsewhere in the USA. One of the striking findings among the Apaches was that meningitis was occurring at a far younger age than was typical. The White River tribal chiefs were under huge pressure because so many babies were dying from "brain fever." Deaths were not the only problem. Many of those who survived had serious problems such as hearing loss, seizures and poor performance at school. Mathu asked me to visit so that we could discuss the current status of research on the *Hi*-b vaccines and talk with the tribal leaders about what might be done.

On arrival in Phoenix, I rented a car to drive the 200 miles to White River, a spectacular drive through the rather intimidating but beautiful Salt River Canyon. There was a magnificent sunset, so I parked my vehicle to enjoy the spectacle — perhaps unwisely as I later learned that there were many wild animals and rattlesnakes en route. The next day, Mathu and I met with the tribal council. In the early 1980s, *Hi*-b conjugate vaccines were at the earliest stages of development (see Chapter 13) and were years away from being implemented in the clinic. John Robbins had shown in laboratory studies that he could improve the immunogenicity of PRP through conjugation. But the crucial studies in human infants had not yet been done.[d,2-4] I was desperate to come up with some way of helping. It seemed an archaic solution, but the only possibility I could think of was the old-fashioned use of passive

---

[d] One of the first breakthrough publications on successful conjugation of *Hi*-b polysaccharide to different proteins and their improved induction of antibodies in mice and rabbits was published in 1980 (see Ref. 2). Porter Anderson and David Smith prepared conjugates (1983) (see Ref. 3) and demonstrated improved induction of antibodies in human infants (1986) (see Ref. 4).

immunisation — infusions of antibodies — an idea appropriate to the early nineteenth-century scientists such as Flexner as I described in Chapter 5. But, since almost all disease in the Native American babies occurred before 12 months,[e] at least this old-fashioned treatment could be a stop-gap while awaiting a vaccine.

Luckily, a colleague, George Siber, who had also been a trainee in the Infectious Diseases Fellowship programme at Boston Children's Hospital in the early 1970s, was now in charge of the provision and distribution of high-quality antibody preparations at the Massachusetts State Laboratories. George apparently had developed a form of treatment called *Bacterial Polysaccharide Immune Globulin (BPIG)* by immunising adult volunteers with bacterial capsular polysaccharides. He then bled these volunteers whose sera contained high antibody concentrations. BPIG was highly effective in preventing *Hi*-b meningitis in the infant rat model and had been used in the clinic to protect immunocompromised children who were at extremely high-risk of life-threatening infections. But BPIG was still considered an experimental product that had not been approved by the Food and Drug Administration (FDA). There were many discussions with the tribal chiefs, following which they would convene to have their own private meetings. The Apache population had often been exploited by researchers who would conduct studies, publish the data and then leave the reservation never to be seen again. But cases of *Hi*-b meningitis continued unabated and, after many months, the tribal chiefs unanimously decided that they wanted to go ahead. Mathu had earned the confidence of the tribal chiefs and they trusted him completely.

A relatively small charity, the Thrasher Research Fund, provided funding and the project was set up in which over 750 babies received either BPIG or *placebo* (saline) at 2, 6 and 10 months of age. We were all quite nervous; what if the treatment didn't work? Worse still, what if there were some serious adverse events? It was a very difficult study because to give the repeated BPIG injections, the medical team had to travel distances of up to 30 miles. There were no cell phones in those days, so no way to check if the family was

---

[e] In the general US population about 20% of cases occurred before 6 months of age compared to 40 to 50% of cases. In the Apache population, 90% of cases occurred before 1 year of age. This was quite a challenge it would mean that any vaccine approach would require prevention of disease in the first 6 months of age.

at home before driving to their house. Nonetheless, Mathu's team surmounted all the hurdles. The study — it took a further three years to complete — showed that BPIG was highly effective in eliminating *Hi*-b meningitis on the reservation.[5] The dreaded "brain fever" was vanquished and Mathu became a celebrity among the grateful Apaches, demonstrating to the world for the first time the importance of preventing meningitis in an economically deprived population. It would be a big stepping stone for Mathu, who over the next two decades would conduct a remarkable series of studies of global importance in the prevention of *Hi*-b meningitis.

Although I made occasional trips to White River, I was of course largely consumed by my responsibilities at Johns Hopkins. Marianne was teaching in the Baltimore School System in charge of the Gifted and Talented Education Programme and on the last day of December 1981, Sarah, our second child was born. As a family, we enjoyed excursions to the Chesapeake Bay and the Eastern Shore of Maryland at weekends. But all this was about to change when, out of the blue, I received a letter from Oxford University.

References

[1] Rubin, L.G., Medeiros, A.A., Yolken, R.H., and Moxon, E.R. Ampicillin treatment failure of apparently beta-lactamase-negative *Haemophilus influenzae* type b meningitis due to novel beta-lactamase. *Lancet*, 1981, 318:1008–1010.

[2] Schneerson, R., Barrera, O., Sutton, A., and Robbins, J.B. Preparation, characterisation and immunogenicity of *Haemophilus influenzae* Type b polysaccharide-protein conjugates. *Journal Experimental Medicine*, 1980, 152:361–376.

[3] Anderson, P. Antibody response to Haemophilus influenzae type b and diptheria toxin induced by conjugate of oligosaccharides of the type b capsule with the non-toxic protein CRM197. *Infection and Immunity*, 1983, 39:233.

[4] Anderson, P., Pichichero, M.E., Insel, R.A., Betts, R., Eby, R., and Smith, D.H. Vaccines consisting of periodate-cleaved oligosaccharides from the capsule of *Haemophilus influenzae* type b coupled to a protein carrier: Structural and temporal requirements for priming in the human infant. *Journal of Immunology*, 1986, 137:1181–1186.

[5] Santosham, M. *et al.* Prevention of Haemophilus influenzae type B infections in high-risk infants treated with bacterial polysaccharide immune globulin. *NEJM*, 1987, 317:923–929.

Reverse Culture Shock and the Dreaming Spires of Oxford

"At a recent meeting of the Electoral Board of the University of Oxford for the vacant post of the Professorship of Paediatrics, your name was put forward ..." ran the letter that went on to ask if I'd be willing to visit in the near future with a view to being a candidate.

This was a bolt from the blue. I had not been a part of the UK medical scene for more than a decade and, anyway, why would I go back to England? It was certainly going to be a major upheaval for Marianne and our two young children. Meantime, I had research funding, a prestigious academic position and a good life in the US. But a part of me was emotionally captured by the idea of Oxford and of what it would offer. For sure, a completely new challenge.

I agreed to what I understood to be a preliminary visit with the aim of exploring the Oxford post with two of the University's most senior medical scientists: David Weatherall and Henry Harris, the Nuffield and Regius Professors of Medicine, respectively. I was not expecting to be interviewed, but that is what happened. In the US, recruitment for a chair would involve a two-day visit, one or two formal lectures, a multitude of meetings with various faculty — but no formal interview.

The morning after my arrival in Oxford, I was summoned to a musty room in the University Offices by a suited functionary. Still jet-lagged, I was almost immune to the gravity of the whole process as I was literally ushered into a room full of *éminences grises*, pompous and rather ridiculous in their academic gowns. I caught sight of David Weatherall with whom I had had a pub supper the night before, who somehow managed to look as if the whole thing was something of a charade. I even imagined that he winked at me, although I'm sure he was discrete.

Formalities were exchanged before the Professor of Paediatrics from Cambridge was asked to open the questions. "Richard and I have met before," he said addressing the Chair, resting his gaze on members of the committee. Avuncular and smiling, he continued, "Indeed, Richard applied for a job in my department many years ago. I didn't hire him — and it was the best decision I ever made!" It wasn't clear what he was driving at and the faces of the electors reflected their puzzlement. "You see," he continued, "I realised he'd be much better off continuing his career in the United States as we had absolutely nothing to offer him in terms of a post that would allow him to continue his research; infectious diseases simply does not exist as a specialty in the UK." And, yes, I *had* indeed briefly entertained the idea of returning to the UK after leaving Boston in 1973, but one short visit immediately after Marianne and I were married was enough to stop that possibility in its tracks. I recall vividly thinking about the stark contrast between the opportunities I was enjoying in the US compared to the UK where the possibilities of continuing clinical training and doing research were virtually non-existent. It was chalk and cheese.

"How would you set about establishing paediatric infectious diseases in the UK if you were to be offered this post?" I was asked. He knew full well that the subspecialty was well established in the US but not at all in the UK. I can't remember how I responded, but he seemed to be on my side, as he'd been a long-standing advocate of the importance of specialist expertise in infectious diseases of childhood. Then Henry Harris, the Regius Professor of Medicine, dug deeply into my research on the genetics of pathogenic bacteria. The serious part of the interview was underway.

"I'm surprised that a paediatrician would be doing research on the genetics of bacteria rather than of human disease," Harris said. "Wouldn't developmental biology be the sort of research that would be more relevant to child health?" This was a question that seemed a gift from on high.

"I'm surprised you're not more concerned about the huge toll of infections on children in all parts of the world and about the role of bacterial diseases as major causes of death and disability," I responded. "One of the neglected areas of medical research in the UK is that of infectious diseases, and my research aims to uncover the fundamental biology of an important disease of children, bacterial meningitis, in order to facilitate better treatment and prevention." Unwilling and unable to adapt to the rather tense atmosphere,

I cheekily asked the electoral board if any of them were aware of progress on meningitis vaccines. Their body language told me that they weren't. Henry Harris had dealt me a card I liked.

An hour or so after the interview, I was told that David Weatherall was anxious to speak to me. I made my way to his office. In his low-key, northern accent, he said, "They've decided to offer you the position, so you're going to have to decide whether you want to take it or not." It was only much later that I came to understand the background to my proposed appointment. Unknown to me, there had been careful, behind the scenes transatlantic communications. The subtle shifting and gliding of the tectonic plates of science and politics had shaped the decision to try and recruit me to Oxford. My name had come forward in the first place because of the strong bond between Henry Harris, the Oxford Regius Professor, and John Littlefield, my Head of Department at Johns Hopkins. Both were eminent cell biologists and Harris had asked Littlefield if he knew of anyone with a background in molecular biology who would be a suitable candidate for the vacant Chair at Oxford. At the same time, Weatherall was pursuing a vision to set up a new research institute in Oxford. His dream was to facilitate the translation of molecular genetics to advance progress in the understanding, treatment and prevention of major diseases. The Medical Research Council had warmed to this idea, but two of its highly influential scientific protagonists, Sydney Brenner and Jim Gowans, had just visited several of the top medical schools in North America and had been hugely impressed by the prominence and research progress in what in the UK was the neglected clinical subspecialty of infectious diseases. It was their opinion that an Institute of Molecular Medicine ought to include expertise in the translation of microbial genetics to help solve the major challenges of infections. The vacant chair offered an opportunity to appoint a professor whose research would tie in with this plan. Weatherall told me that the Medical Research Council were very keen on strengthening infectious diseases and that I would be in a very strong position to get substantial research funding. Recruitment talk of course, but I knew in my heart of hearts that I would have great difficulty in turning the opportunity down, although I told him I needed more time to come to a decision. It was indeed a huge step for us as a family. Marianne and I decided that I should request a further visit to allow us to explore together in much more detail what Oxford would offer.

I anticipated that there would be tough negotiations on the terms of my appointment. In fact, I quickly found out how differently Oxford University approached academic appointments compared to what would have happened in the US. In North America, one prepared a shopping list of essential requirements — space, essential personnel, laboratory equipment and start-up funds — which then became the basis of a prolonged bartering process with the Medical School Dean, culminating in a "who blinks first" negotiation. Oxford Medical School did not and still does not have a Dean, so my meeting was with the Regius Professor, Henry Harris, by whom I had been grilled at the interview. As I began outlining the highest priorities on my "shopping list," he interrupted me with a knowing smile, "Look Moxon," he said, "you're taking a typically American approach and we don't play that game. I won't be able to promise anything until you agree to take the position. Then I will work as hard as I can to get you what you need." I was completely taken aback. I had come meticulously prepared for a dialogue and tough negotiations, but these were apparently simply not going to happen.

"But," I protested, "how can I sign up to something as important as a Chair at Oxford University without any reassurances that I will get at least the essential support for my research?" Harris responded with a calm, almost patronising air, his tone verging on the impatient: "We are appointing you to a very important position and I am aware of what it will take for you to be successful in your research. Moxon: think this through; why would we want you to fail? It doesn't make any sense to deny you things that are a *sine qua non* for you to be able to do your research. However, I can't do anything until you agree to come. OK, if you accept and it doesn't work out, although I think it will, your appointment here would be a ticket for you to go anywhere you want to go. Oxford is Oxford!"

It was an unexpected and peculiar feeling to have my arm twisted in this fashion. "I tell you what," Harris continued, "if we can shake hands on the fact that you are coming, then by the end of the day, I will talk with the great and the good and get them to agree to fund two of the research posts that you want and guarantee the funds for the refurbishment of your laboratory. Consider this an indication of my good will and ability to deliver promptly. We are serious about you coming, so if you agree, we can go from there. I can't promise more." So, that was that. By the end of the afternoon, Harris had duly delivered on his immediate promise and I was given terrific support over the next years, in large part through his and David Weatherall's influence.

A year later, I began what would be 25 years as the Chair of Paediatrics in the Oxford University medical school. At Johns Hopkins, I had headed up a unit with four internationally recognised specialists in paediatric infectious diseases. Here in the UK, I was at the time the only children's infectious disease specialist in the country. A few months earlier, Bill Marshall had died suddenly and prematurely (aged 54). Bill, appointed in 1977 as the infectious diseases specialist at Great Ormond Street (GOS), was the brilliant paediatrician who had made such a deep impression on me (see Chapter 4) when I had been a junior doctor at GOS looking after a little boy with *Hi*-b meningitis. It seemed a strange but fitting coincidence that 15 years after my memorable encounter with him, the mantle of Elijah had fallen on my shoulders.

Indeed, it was Bill (who knew about my impending appointment before he died) who had put me in touch with a young paediatrician (David Isaacs) who wanted to pursue a career in infectious diseases and had joined me in Oxford when I arrived to take up my appointment on April 1, 1984. David's enthusiasm and aptitude as a clinician were one of the reasons why any lingering doubts on the wisdom of having made the transatlantic move were quickly dispelled.

In taking on my academic leadership role in Oxford, I needed to combine fundamental research with its translation into the clinic. David and I quickly initiated several projects focussing on infections in the intensive care unit for very sick newborns. It proved a highly successful means to combine the provision of clinical expertise while carrying out some much-needed research on how best to manage neonatal infections. Later, we jointly published a book[2] that, thanks to David's hard work (he did more than the lion's share of the writing), provided a much-used and successful text for neonatologists.

Of course, as I describe later (Chapter 16), I was also frantically working to establish my laboratory research. In 1984, I had no doubt that the *Hi*-b vaccine would eventually become a reality, although there was practically no awareness of this innovative research in the UK even among experts in public health and microbiology. Soon after I started, I had a meeting with the head of the Clinical Microbiology Laboratory at the John Radcliffe Hospital, the main clinical base for the Oxford Medical School, who greeted me with a provocative, thinly concealed scepticism over my research interest. "I know *Hi*-b meningitis is a big problem in the States, but here in the UK, it's nothing like as important," he told me. I wondered what lay behind this challenging premise from Joseph Selkon, an experienced clinical microbiologist with whom I was hoping to build bridges. He referred to an inner-city study from Birmingham, UK, showing that about one in two thousand babies came down with Hi-b meningitis before they were 5 years old, rates of infection that were fivefold lower in North America where the comparable figure was one child out of every 400. I thought it likely that this was a serious underestimation of the burden of infections, but research on Hi-b in the UK had attracted little attention and the prevailing perspective was reflected in a 1967 monograph which concluded that "widespread immunisation, even if satisfactory methods are worked out, is not likely to be considered advisable ..."[1] UK Microbiologists and public health scientists were generally oblivious to the efforts of US researchers on Hi-b meningitis vaccines. As the visit of the UK Medical Research Scientists had uncovered, the US and the UK were parallel universes — more than an ocean apart when it came to clinical expertise and research in infectious diseases.

Knowing that the US trials in humans on the polysaccharide–protein conjugates were imminent, I wanted to be ready to capitalise on this progress so that the UK would be at the forefront of meningitis research in Europe. It was therefore vital to have accurate information on Hi-b epidemiology, the scientific discipline that deals with how often diseases occur, who is affected, at what age, in what geographical locations and with what impact on health.

Cases of meningitis, including Hi-b, are a notifiable disease in the UK; medical doctors have a statutory duty to report each case to the Communicable Diseases Surveillance Centre based in Colindale, near London. But such a system, called *passive surveillance*, is a notoriously unreliable process. *Active surveillance* was needed and could be achieved through national networking.[a]

---

[a] The UK Public Health Laboratory Surveillance (PHLS) had been established through the National Health Service Act in 1946, largely because of fears of bacteriological warfare during World War II. In 1977, the Communicable Disease Surveillance Centre (CDSC) was established within the PHLS. One of its senior microbiologists was Mary Slack, formerly Head of the World Health Organisation (WHO) Collaborating Centre and Global Reference Laboratory for *H. influenzae*. She has worked extensively in developing countries providing technical support, advice and training for sentinel site surveillance of paediatric bacterial meningitis, pneumonia and sepsis. The Oxfordshire public health team was led by Richard Mayon-White, a brilliant infectious diseases epidemiologist and, in his spare time, a renowned expert on the Thames River. A young paediatrician, Gareth Tudor-Williams, was keen to be involved in the project.

Indeed, we were well-placed in Oxford to carry out a comprehensive, prospective study to document the impact of *Hi*-b in the UK. The key was to put in place a mechanism to ensure that every time any microbiology laboratory in the UK cultured *Hi*-b bacteria from a sick patient, it would trigger the collection of essential information to document the rates of meningitis. It was also important to verify, using state-of-the-art DNA tests (developed in my Oxford research laboratory), that each isolate was indeed *Hi*-b and that there had not been a misidentification. This was not a short-term project; we needed to collect data for several years to obtain a reliable set of information. Meantime, I was keeping closely in touch with the progress being made in the US on the *Hi*-b conjugate vaccines and making sure that my new laboratory was getting to grips with the challenges of bacterial genetics, as I describe in the following two chapters.

Reference

[1] *Haemophilus influenzae*. Its Clinical Importance. DC Turk. 1967. English Universities Press, p. 38.

He would later become an international expert on childhood AIDS, although the first case of paediatric Human Immunodeficiency Virus (HIV) infection was not recognised until 1983.

Developing and Implementing the *Hi*-b Conjugate Vaccines

Around 1980, based on the earlier Rockefeller research, Robbins and Schneerson had worked out the required chemical methods to attach the *Hi*-b capsular polysaccharide to a protein. They used the tetanus toxoid vaccine component as the so-called *carrier protein* as it was easily available and approved for human use. When laboratory animals were immunised with this conjugate, it drastically improved the quality and quantity of antibodies compared to the polysaccharide alone.[1] Shortly afterwards, Porter Anderson used somewhat different chemistry to couple the *Hi*-b capsular polysaccharide to a different protein, modified diphtheria toxin,[2,3] vaccine and obtained antibody responses in baby rabbits that exceeded by 100-fold the amount required to protect against meningitis. In both cases, the reason for this striking improvement of the conjugates, compared to polysaccharide on its own, was the activation of an arm of the immune system known as *T-cells* that are critical for the efficient induction of antibody responses (see Figure 13.1). Indeed, on the very day (April 1, 1984) that I started my new job in Oxford, scientists at an international meeting[a] in the US reported the results of the first clinical trial of a *Hi*-b conjugate in infants.[4]

Meantime, in the US, the Rochester scientists had also obtained their first encouraging data on immunising humans. Based on laboratory data, the amounts of antibodies were judged sufficient to protect infants. But now came the monumental challenge of carrying out the large-scale clinical trials on thousands of children. Compelling data were required to satisfy regulatory bodies on the safety and effectiveness of the vaccine. Partnership between the

---

[a] The annual meeting of the American Society for Pediatric Research.

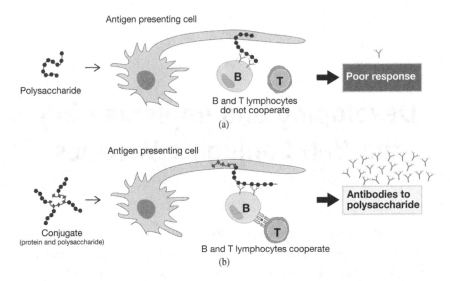

Figure 13.1 (a) When polysaccharide alone is used to immunise animals, it is transported by an antigen presenting cell to local *lymph nodes*, but the amount and quality of antibodies produced are inadequate, especially in very young children, because polysaccharide antigens fail to induce efficient cooperation between B and T lymphocytes. This is called T-cell independence. (b) In contrast, when the polysaccharide is chemically coupled to a protein, there is active cooperation between B and T lymphocytes and the quantity and quality of antibodies is greatly enhanced. The coupling to protein converts the immune response to polysaccharide from being T-cell independent to T-cell dependent.

academic researchers and vaccine manufacturers was needed, but this was complicated. Conjugates were an entirely new concept in vaccine development in the early 1980s, an era when even the production of existing licensed vaccines was proving a challenge. For example, an alleged role of pertussis vaccine in causing brain damage[5] (later shown to be false) had resulted in more than 200 lawsuits in US courts alone,[b] threatening the commercial viability of the manufacturers.[6] The boardrooms of large pharmaceutical companies were questioning their continuing commitment to make routine licensed vaccines, let alone designating funds for research and development of new products.

David Smith had failed to persuade any of the vaccine manufacturers, large or small, to be partners in commercialising the *Hi*-b conjugates made in Porter Anderson's laboratory. John Robbins and Rachel Schneerson — the

---

[b] The average amount of compensation was reported to be around $26 million.

scientists who had led the way on the conjugation chemistry — were in an even trickier situation. They had been unable to get the National Institutes of Health (NIH) patent office to protect their conjugation technology without which no commercial company would even begin negotiations with them. At the time, the NIH had only one patent lawyer, who was not interested in vaccines. The NIH and Rochester scientists, the front-line competitors in the race to get approval (licensure) of *Hi*-b conjugate vaccines, were apparently both stymied.

A solution to the conundrum came from the vision and entrepreneurial skills of David Smith. Renowned for his tough, articulate but often uncompromising manner, the reluctance of vaccine manufacturers to commercialise the *Hi*-b vaccine was unacceptable to him. It ran counter to everything that had been instilled into him by his devoted parents, both teachers, who believed passionately in sticking to your convictions and never giving up. Stung by the resistance of the vaccine manufacturers, Smith resigned his academic post as Chair of Paediatrics in 1983. In what seemed to his colleagues to be a reckless decision, Smith founded a small vaccine development company, Praxis Biologicals. Nothing had prepared him for becoming a businessman, so it seemed on the face of it to be a preposterous gamble. He assembled a small scientific advisory board and outlined a business plan to manufacture *Hi*-b and other vaccines.

Although the plain (unconjugated) *Hi*-b polysaccharide vaccine was ineffective in preventing meningitis in very young children, it did protect children two years or older. So, Smith argued, it could reduce the number of episodes of meningitis by around a third. As a public health policy, the proposal was highly controversial. But there was a societal issue involving day care centres that provided impetus for Smith's proposal. In the 1980s, under the Reagan administration, federal child-care funding for middle and high-income families had nearly doubled, stimulating an enormous growth in profitable day care centres for pre-school children. The close contact between children increased the risk of *Hi*-b meningitis by fivefold causing small outbreaks that were highly publicised. Smith reasoned that the relatively affluent parents who could afford day care would also buy the vaccine. Having their child come down with meningitis was one of their biggest fears.

His idea was supported by a cost-effectiveness analysis carried out independently by the US Centre for Disease Control.[7] Smith kept plugging away and eventually his advocacy prevailed; the plain polysaccharide vaccine was licensed for use in 1983 and marketed in the US in 1985. Using the revenues from sales of the vaccine, along with venture capital funding and his own personal money,[c] Praxis acquired state-of-the-art laboratories in Rochester, New York, and built one of the finest manufacturing facilities in North America in Sanford, N. Carolina. In 1988, the company was given the go-ahead by the Food and Drug Administration (FDA) to begin manufacturing operations at its new 116,000 square foot biologics facility. But, in a major setback, sales were much lower than hoped for and the expenses of the new plant plunged Praxis's finances into crisis. Their revenues declined more than 50% — a $12 million net loss. There was no money to pay the 40 or so scientific staff and the Praxis Board felt obliged to pass a vote of no confidence in Smith and recommend the sale of Praxis. Smith had other ideas. He dismissed the entire board, borrowed more money and, in a remarkable feat of organisation, orchestrated a trial on the Praxis *Hi*-b conjugate vaccine in 60,000 children. This showed that the vaccine was 100% effective in preventing *Hi*-b meningitis.[8]

This was crucial evidence with which to toughen negotiations with the FDA who issued a licence for the conjugate vaccine in December 1987. Smith's small company of less than 50 staff had won a tense, often acrimonious race against the much larger pharmaceutical company Merck Sharp and Dohme (MSD) who had developed a rival *Hi*-b conjugate vaccine. MSD had more than 10,000 employees involved in making vaccines. Their *Hi*-b conjugate vaccine was licensed by FDA soon after the Praxis vaccine, following a successful trial demonstrating protection of Apaches in Arizona conducted by my former trainee, Mathu Santosham. As described in Chapter 11, the passive immunisation of Apache infants with antibodies was a temporary intervention pending the development of *Hi*-b conjugates. I was on the scientific advisory board for the trial and had worked closely with Mathu on it. What followed was a fierce bidding war as MSD attempted to acquire Praxis Biologicals, their main rival, whose conjugate vaccine had already been licensed. But, in June 1989, Praxis was acquired by American

---

[c] In the documentary *The David Hamilton Smith Story*, one of his daughters states that he remortgaged his house to raise capital. Available at https://vimeo.com/72789961 (Accessed 22 April 2021).

Cyanamid, a big pharmaceutical company that had the financial backing and manufacturing capacity to produce the millions of vaccine doses required for routine immunisation. A jubilant Smith, exhausted after months of working 18 hours a day, could now enjoy time with his family and think about what to do with his life after netting some $90 million[d] through the sale of Praxis.

By 1990, three *Hi*-b conjugate vaccines had been approved by the FDA for use in infants and children (2 months or older). The impact of each of the vaccines was spectacular. Within months, *Hi*-b meningitis, as well as the other serious diseases[e] caused by this bacterium, decreased by more than 95%. *Hi*-b conjugate vaccines were safe and highly effective, a milestone in public health for which, in 1995, Robbins, Schneerson, Smith and Anderson were awarded the prestigious *Lasker Award*. The coupling of *Hi*-b polysaccharide to a carrier protein provided a means to overcome the naturally poor immune responses to polysaccharides in the very young, those individuals most in need of protection against the deaths and disabilities caused by *Hi*-b meningitis. The same principles would later prove equally applicable to prevention of meningococcal and pneumococcal meningitis whose polysaccharides, unless conjugated, were ineffective as vaccines in the very young.

In the UK, by the time that conjugate vaccines were being licensed and implemented in the US in 1989–1990, our data (five years of surveillance) showed that one in 850 children became sick with *Hi*-b meningitis before their fifth birthday. This made it the major cause of bacterial meningitis in the UK. The public health impact was not just the large number of deaths. With treatment, more than 90% of those who got *Hi*-b meningitis survived. But of those who recovered, ten percent had major disabilities such as convulsions, deafness, visual loss and learning problems. The humanitarian and economic costs of this accumulating population of brain damaged children from *Hi*-b meningitis was a much more serious public health problem than many UK

---

[d] Much of David Smith's remaining life was devoted to philanthropic causes. He died prematurely of malignant melanoma in 1999.

[e] Although 70% of *Hi*-b invasive infections are meningitis, the bacterium also causes septicaemia, bacterial croup (epiglottitis), septic arthritis, cellulitis and pneumonia.

experts, including my Oxford microbiologist colleague, had conceded. There was a compelling case for implementing routine *Hi*-b immunisation.[f]

Trials were needed because of differences in the timing of the routine immunisations given to infants in the UK. For example, the US gave routine infant immunisations at 2, 4 and 6 months, whereas the UK schedule was 3, 5 and 9 months. I put together a formal application to the Medical Research Council (MRC) for funds to carry out this study. It was turned down, the reviewers considering it of insufficient priority given the huge competition for research funding. Besides, MRC argued, this kind of applied research ought to be funded by vaccine manufacturers. The issue of who should be responsible for funding the research required to implement new vaccines remains a thorny and complex issue of huge public health importance. An appreciation of what must be done (and spent) to ensure safe and effective vaccines is poorly understood by most people, although public awareness of these issues has been heightened recently by Ebola and COVID-19.

There is a four-stage process that is required to pave the way for the routine implementation of most vaccines. In the pre-clinical stage of testing, researchers give the vaccine to animals to see if it is tolerated and triggers an immune response. In phase 1 of clinical testing, the vaccine is given to a small group of people to determine whether it is safe and to learn more about the immune response it simulates. The vaccine is then given to hundreds of people (phase 2) so scientists can learn more about its safety and correct dosage. Finally, in phase 3, the vaccine is given to thousands of people to determine if it prevents infection and to gather more data on safety, usually through comparison with a control group which is given a placebo.

To investigate the immune responses to the *Hi*-b conjugate vaccine in Oxford children, a phase 2 trial was required. Because of the lack of external financial support, the project depended on the good will and cooperation of general practitioners, the regional public health infrastructure and, very

---

[f] It is a good example of how perspectives on public health issues can change in a relatively short time frame. In less than a decade, *Hi*-b conjugate vaccines would become a routine immunisation in the UK and many other countries and, within two decades, were integrated into the World Health Organisation's global Expanded Programme in Immunisation.

importantly, the enthusiasm of parents. Because meningitis strikes rapidly and is life-threatening, it consistently ranks high on the list of diseases for which the public want a vaccine. Indeed, even when the Hi-b vaccine was still in the research phase, parents were enthusiastic for their children to be immunised and this made it easier to enlist the cooperation of parents and the support of their GPs.

There was also strong encouragement from the members of a Medical Research Council sub-committee of which I was the chairman. This included an observer from the Department of Health, David Salisbury. Recently appointed to oversee UK immunisation activities and policies, David was already making a strong impact through his astute handling of the routine infant vaccine programme. He organised district by district meetings holding to account more than 200 public health officers throughout England and Wales — many of whom found these encounters terrifying. But leadership and action were badly needed as the uptake of routine immunisations in the UK in the 1980s was strikingly lower than other comparable European countries. A scathing report[9] had concluded that low immunisation uptake was a consequence of "...poor administration, lack of professional commitment and inadequate support for those providing the service." Less than 70% of infants had received the measles vaccine, in part an indirect knock-on effect of mistrust in the whooping cough (pertussis) vaccine because of widespread but unfounded fear (as mentioned on page 114) that it might cause brain damage.[g]

Our small trial showed that the Hi-b conjugate vaccine induced protective levels of antibodies in UK infants, encouraging results that I presented to the MRC committee prior to its formal publication. What followed was a nasty surprise. David Salisbury informed us that, following months of discussion within the Department of Health, a radical change in the timing of the routine UK immunisation schedule was imminent. The Hi-b immunogenicity study, performed on the "old" schedule, needed to be re-done using the new 2, 3 and 4-month-old schedule.[h]

---

[g] It had resulted in a nationwide epidemic of whooping cough in the UK (1989) in which 5,000 children were admitted to hospital, many with devastating complications such as pneumonia and convulsions. This had a ripple effect on trust in other vaccines.

[h] In Chapter 16, I describe the problem of overlap and communication between research in the Department of Health and academia. The whole idea of having a Department of Health

At the time, there had been no completed[i,10] human trials to show the efficacy of the vaccine using the new UK immunisation schedule. Luckily, a new Australian trainee arrived in my department at just the right time to spearhead a small phase 2 study of a vaccine (known as PRP-T) made by Pasteur Merieux Connaught in Oxford children using the new immunisation schedule. I probably underestimated how challenging these projects were for these young trainee paediatricians, many of whom had had little if any experience of clinical research. My new trainee didn't hesitate to let me know that he felt too much was expected of him, but he quickly got the project underway and proved more than equal to the task. The results were encouraging and made possible a much more ambitious study: an evaluation of the protective efficacy of PRP-T in Oxfordshire children.

While this was in the planning stage, momentum for a proposal to include *Hi*-b conjugate vaccines in the routine UK immunisation schedule had gathered pace. After a one-day workshop that I had organised in Oxford in 1990, there had been a virtually unanimous consensus in favour of national implementation of the *Hi*-b vaccine. This was followed almost immediately by an emergency meeting of the UK Joint Committee on Vaccines and Immunisation (JCVI[j]), the government's advisory body, who formally recommended its introduction. The Department of Health set the start date for October 1992 — enough time to complete the large PRP-T efficacy study, but only just.

I assembled a project team to communicate our proposal to the large number of participating general practices in the Oxford region and obtain their cooperation. Some serious concerns were raised about the extra work and financial implications for GPs, health visitors and immunisation clinics. The Minister of Health, Kenneth Clark at the time, had set up a scheme whereby GPs only got remuneration for giving routine immunisations if they achieved uptake targets of greater than 90%. The addition of the *Hi*-b conjugate vaccine, to be given at 2, 3 and 4 months, was a potential complication. Each

---

'observer' on the committee was to facilitate good communication. I was disappointed that information about the change in the timing of the immunisation schedule had not been shared with the committee.

[i] Results from two prematurely terminated controlled studies, uncontrolled estimates and laboratory assays showing protective activity carried out by Pasteur Merieux were published in 1992 (see Ref. 10).

[j] Not to be confused with the other JCVI (J. Craig Venter Institute).

child would need to receive an additional injection and resources were needed to meet the responsibility of explaining what was involved to parents and to train personnel in giving the extra vaccine. There were concerns about the possibility of adverse events due to the vaccine and a substantial investment of time and diplomacy was needed to garner the full support of GPs, although the majority were positive about the prospect of protecting infants against the much-feared *Hi*-b meningitis.

Once again, my application for research support from the Medical Research Council was unsuccessful. The expert reviews were highly critical; they wanted a classical, randomised controlled trial. But this would have been prohibitively costly and, given the compelling success of the vaccine elsewhere, arguably unethical because those children randomised to the comparison group would be denied the benefits of an effective vaccine. Fortunately, Pasteur Merieux Connaught was enthusiastic about the trial and agreed to provide funding, although there were concerns that, as an academic research unit, Oxford should not be collaborating with a commercial company that could profit from future sales of the vaccine, an important issue that I discuss further in Chapter 16.

There were eight districts in Oxfordshire and a fortuitous logistical issue determined the design of the trial. Before a child can be immunised in the UK, parents must present a card generated by a health district computer. This card is used as the basis for documenting parents' assent to immunisation. But the computer systems were so antiquated that only four of the eight districts could be re-programmed to add the additional *Hi*-b immunisation. Although not the randomised, double blinded trial demanded by the purists, the enforced circumstances fortuitously created a potentially satisfactory and inexpensive trial design. About half the babies would be immunised with the *Hi*-b conjugate vaccine and half (acting as a comparison group) would not receive the vaccine.

By 1991, four different *Hi*-b conjugate vaccines had been licensed of which one (known as PRP-D and manufactured by Connaught) was withdrawn as it had proved disappointingly ineffective in preventing infection in a high-risk population of babies in Alaska.[11] The vaccine produced by Merck Sharp and Dohme had also run into problems because of variability in the manufacturing process and some batches of the vaccine had to be recalled. Of the two remaining vaccines, only the Praxis vaccine (American Cyanamid had been acquired by Wyeth Lederle) had been shown to protect against disease in a large human trial. It seemed that there would be an ethical imperative to favour its use over PRP-T, which lacked clinical data on protection. The situation was yet more complex. There was insufficient vaccine to provide the millions of doses that would be required to implement the vaccine that was scheduled to begin in October 1992. All of these factors taken together meant that a trial was needed to show the efficacy of PRP-T.

Was there time to complete this trial within this 15-month window? Calculations showed that enrolment of almost 11,000 Oxfordshire children (half to be offered routine immunisations and *Hi*-b vaccine, the other half only the routine vaccines) would be equivalent to 12,000 years of child exposure to *Hi*-b. If the vaccine was highly effective (~90% protection), there ought to be no more than one case among immunised children compared to an expected 10 cases in the unimmunised cohort. If we were to have results, then the first infants needed to be immunised by May 1991. There was little time to lose.

References

[1] Schneerson, R., Robbins, J.B., Barrera, O., Sutton, A., Habig, W.B., Hardegree, M.C., and Chaimovich, J. Haemophilus influenzae type B polysaccharide-protein conjugates: Model for a new generation of capsular polysaccharide vaccines. *Prog Clin Biol Res*, 1980, 47:77–94.

[2] Pappenheimer, A.M., Uchida, T., and Harper, A.A. An immunological study of the diphtheria toxin molecule. *Immunochemistry*, 1972, 9:891–906.

[3] Anderson, P., Pichichero, M.E., and Insel, R.A. Immunogens consisting of oligosaccharides from the capsule of Haemophilus influenzae type b coupled to diphtheria toxoid or the toxin protein CRM197. *Journal of Clinical Investigation*, 1985, 76:52–59.

[4] Ward, J., Berkowitz, C., Pescetti, J., Burkhard, K., Samuelson, O., and Gordon, L. Enhanced immunogenicity in young infants of a new Haemophilus influenzae type b (Hib) capsular polysaccharide (PRP)-diphtheria toxoid conjugate vaccine Enhanced immunogenicity in young infants of a new Haemophilus influenzae type b (Hib) capsular polysaccharide (PRP)-diphtheria toxoid conjugate vaccine. *Pediatric Research*, 1984, 18:287.

[5] Kulenkampff, M., Schwartzman, J.S., and Wilson, J. Neurological complications of pertussis inoculation. *Archives of Disease in Childhood*, 1974, 49:46–49.

[6] Koplan, J.P. and Hinman, A.H. Decision analysis, public policy and pertussis: Are they compatible? *Medical Decision Making*, 1987, 7:72–73.

[7] Makintubee, S., Istre, G., and Ward, J.I. Transmission of invasive *Haemophilus influenzae* type b disease in day care settings. *Journal of Pediatrics*, 1987, 111:180–186.

[8] Black, S.B., Shinefield, H.R., Fireman, B. *et al.* Efficacy in infancy of oligosaccharide conjugate Haemophilus influenzae type b (HbOC) vaccine in a United States population of 61,080 children. *Pediatric Infectious Disease Journal*, 1991, 10:97–110.

[9] Nicholl, A., Elliman, D., and Begg, N.T. Immunisation: Causes of failure and strategies and tactics for success. *BMJ*, 1988, 299:808–811.

[10] Fritzell, B. and Plotkin, S. Efficacy and safety of a Haemophilus influenzae type b capsular polysaccharide-tetanus protein conjugate vaccine. *Journal of Pediatrics*, 1992, 121:355–362.

[11] Ward, J.L. and the Alaska Study Group. Limited efficacy of a haemophilus influenzae type b conjugate vaccine in Alaska native infants. *NEJM*, 1990, 323:1393–1401.

Laboratory Research: Bacterial Genetics in Oxford

Clinician-scientists lead a somewhat frenetic existence, wearing several hats so that most days are a complex juggling act between taking care of the sick, teaching and research. While orchestrating the *Hi*-b epidemiology and vaccine trials, my research laboratory was trying to make further progress on the genetics of the type b capsular polysaccharide.

I assigned the project to a trainee in my laboratory, a former Oxford University music scholar who had studied physical chemistry as an undergraduate before training in medicine. Simon Kroll brought to my mind a youthful Samuel Pickwick (Charles Dickens); round-faced, clean shaven, bespectacled, inquisitive and boyishly enthusiastic. His intelligence and undisguised ambition appealed to me from our first meeting. This was his first foray into basic research since his time as an undergraduate, but thanks to the guidance of some technically experienced scientists in my group, Simon's apprenticeship went well. He was a fast learner, worked tirelessly and within a few months was proficient in using the tools of molecular biology. Over the next couple of years, Simon mapped the genes required for the transport, assembly and synthesis of *Hi*-b polysaccharide and then, using other strains, he mapped the genetic regions for the five other capsular types.[a] It was a *tour de force* that allowed us to do detailed comparative virulence studies. It led to an exciting collaboration with US colleagues who had developed a technique to index the genetic variations among the hundreds of different isolates of *Hi*, including the many different capsular types. Their research assigned a

---

[a] See Chapter 3, Ref. 12.

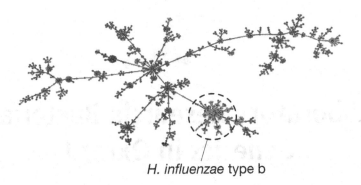

H. influenzae type b

Figure 14.1   Phylogenetic tree of a diverse collection of clinical isolates of *H. influenzae*. All the different capsular types and unencapsulated strains of *H. influenzae* are represented (only those of capsular type b are indicated). Each dot represents the DNA profile (genotype) of one clinical isolate. Note the clustering of type b isolates of similar genotype. Adapted from later, more detailed studies of the population biology of *H. influenzae* (see Ref. 1).

profile, analogous to a bar-code, to each isolate and computer software then generated a visual representation (called a *phylogenetic tree*) to depict the genetic variations (Figure 14.1).

This analysis had thrown up an important finding: isolates of each of the six distinct capsular types had virtually the same bar-code and appeared on the phylogenetic tree as discrete clusters or genotypes. The genetic similarity of isolates of each capsular type raised an important question: was the heightened virulence of *Hi*-b determined by its capsule or by other genes that were unique to that genotype? Using a racing car analogy, would transferring the engine (genes for the b capsular polysaccharide) into the chassis (genome) of a different *Hi* type make it more powerful (virulent)?

Using DNA transformation (see Figure 10.1), six *Hi* variants were constructed, each differing from the others uniquely in their capsule genes. Different engines, same chassis. These were then compared for their ability to cause meningitis using our infant rat experimental model. The results were decisive. Only the type b transformants caused meningitis. The numbers of bacteria in the blood were a thousand-fold greater than any of the other capsular types. The primacy of the type b polysaccharide in determining enhanced virulence provided reassurance that if a switch in capsular type did occur in nature, then variants lacking it would be drastically reduced in their virulence and the effectiveness of the vaccine would not be substantially compromised. The completion of this successful research was a relief. On many occasions,

I had lain awake at night, fearing that in the UK I would not be able to achieve the calibre of research that I had accomplished in the US.

In my last year at Johns Hopkins, our research had shown that alterations in a different *Hi*-b surface structure, endotoxin, could also affect virulence. Prior collaboration with one of my US colleagues had provided an antibody that specifically reacted with the *Hi*-b endotoxin; I thought we could apply the same recombinant DNA approach that I had used to isolate genes for capsule synthesis to identify the genes for making endotoxin.

There was another incentive for these experiments. Some bacteria reacted strongly with the antibody and others did not. But when the colonies were re-grown, positive colonies could become negative and *vice versa*. We hadn't a clue as to what caused this switching and whether it had any relevance to the infectious process. The idea was to identify fragments of DNA that could restore antibody reactivity to a strain of *Hi*-b bacteria that was unreactive. This was an ideal project for a graduate of Harvard Medical School, Jeff Weiser, who had recently joined my laboratory. For many months, Jeff encountered one frustration after another; just when success seemed around the corner, repeat experiments failed to confirm the result. How long do you go on before abandoning an approach that is not producing consistent results when there is no guarantee of success? Most ideas and experiments in science just do not work out, not necessarily because the idea is problematical but, frustratingly, because the available methodology is not capable of delivering the goods.

We were nearing the point of giving up when a DNA transformation experiment showed a small percentage of colonies that reacted with the antibody and, in contrast to previous false dawns, the result was reproducible. From the library of genomic fragments, Jeff had found a fragment of DNA that could transform *Hi*-b bacteria so that these negatively reacting colonies became positive. This was exciting, but there was a further unexpected and thrilling finding. The DNA sequence of the cloned DNA was highly unusual, in that there were multiple (20) repetitions of four nucleotides.[b] Short repeats of this kind were at the time a well-known characteristic of human DNA (they are called *microsatellites*) but not of bacterial genomes. The DNA

---

[b] Cytosine, two adenosines and thymidine or, as conventionally abbreviated, CAAT.

repeats explained the variable reactivity of the *Hi*-b endotoxin molecule with the antibody, a topic to which I return later.[c] It would prove one of the most exciting discoveries from my laboratory, although, as is so often the case, its wider significance as a major mechanism of bacterial surface variation was not at all apparent to me at the time.

The result was timely as I was about to apply for a renewal of my Medical Research Council funding. The imperative of getting research support is something of a nightmare for scientists as without it years of work can literally be brought to an abrupt halt. Scientists know just what a gut-wrenching, hugely competitive challenge it is — the cause of much anxiety and lost sleep. In the five years since I had started in Oxford, the number of people in the laboratory had increased, with a commensurate need to obtain support for salaries, equipment and laboratory reagents. My start-up funding provided by the University was spent, but I had acquired some additional financial support through a donation from the National Meningitis Trust. This came about through an extraordinary circumstance.

Starting in 1981, two smallish towns in Gloucestershire (Stonehouse and Stroud), located in the heart of the Cotswolds in the south of England, had experienced an outbreak of meningitis caused by the meningococcus B variant (MenB) for which there was no vaccine. Over a period of 5 years, there were 65 cases, about 10 times the expected number. It is still unclear why this outbreak occurred in this rural community, but it became headline news. It affected the local housing market and impacted tourism to this picturesque part of the Cotswolds. Health experts were baffled and there were hastily arranged public meetings. The local community was shocked and terrified. In the absence of an explanation for the outbreak, it was variously attributed to a new power station, a public swimming pool and even a milk factory. Fear turned into anger, even malice; the parents of a child who was admitted to hospital with meningitis found a skull and cross bones affixed to their door. The Parliamentary Under-Secretary of State at the Department of Health and Social Security in the Thatcher Government[d] was summoned to visit and report back to the House of Lords.

---

[c] See Chapter 17.

[d] Baroness Trumpington, a former code breaker at Bletchley Park.

Among the victims who died from meningococcal infection in the Stroud MenB outbreak was the son, aged 11 months, of the owner of a highly successful truck haulage business. Devastated by the loss, Steve Dayman and his wife used their own and a close friend's personal money to set up the National Meningitis Trust, a charity that aimed to raise awareness about this devastating disease and to provide support for affected families. It also recognised the need for further research and one of Steve Dayman's co-workers, whose 2-year-old son had also suffered, but recovered, from meningitis, had told him about my research programme in Oxford. A visit to my laboratory was arranged and I was grilled about the prospects of a vaccine. I explained that there were special research challenges concerning MenB[e] and that it was unlikely that there would be a vaccine for many years.

Some weeks later, I learned that the Trust had awarded my laboratory substantial research funding, a huge incentive to take on the challenge of research into developing a MenB vaccine. But to make a proper fist of this, I knew that renewal of my Medical Research Council Research Programme was vital. These were turbulent times for biomedical research funding. In the mid-1980s, Margaret Thatcher had been at odds with her Whitehall colleagues over policies for funding scientific research. In a transformative political speech, her predecessor, the Labour Prime Minister Harold Wilson, had proclaimed more than two decades earlier that many in government were Luddites, ignorant of science and unfit to embrace the opportunities that were essential for a strong economy. If the UK was to prosper, he argued, "Britain must seize the opportunity for a scientific revolution based on the white heat of technology ... the cloth cap must be replaced by the white laboratory coat." Thatcher, who had a degree in Chemistry from Oxford University, was a passionate advocate of science and saw basic research as an investment that would have general application even if, as historically was so often the case, its benefits were unforeseen by their investigators. This curiosity-driven, so-called "blue-skies" research, was different from the kind of applied research of my laboratory. My aim, in line with Weatherall's vision for the Institute of Molecular Medicine, was to use these new discoveries to improve clinical practice, often called *translational research*.

---

[e] See Chapter 15.

The MRC review process of scientific research applications pays a lot of attention to the applicant's track record of scientific productivity. On the plus side, my laboratory had published several high quality original scientific papers. But there had to be more thrust to my proposal than simply a continuation of our *Hi*-b research. An MRC Research Programme is defined as one that: "aims to help the medical science community think bigger through a coordinated and coherent group of related projects, which may be developed to address an inter-related set of questions across a broad scientific area." So, it was opportune to widen my horizons to include research into the meningococcus. The imminent introduction of the conjugate vaccines meant that continuing research into *Hi*-b meningitis was likely to be given a lower priority by MRC reviewers.

Because of its broad implications for vaccine development, the discovery of DNA repeats (mentioned above) as a driver of variable expression of bacterial surface structures resulted in an invitation for me to speak at an international conference on the meningococcus[f] held in Berlin in 1990. For the best part of a week I was plunged into cutting-edge research into the meningococcus, including the most up-to-date research on a MenB vaccine. In choosing endotoxin as a candidate meningococcal vaccine antigen, I had to consider the importance of the mechanism driving its variable expression — a potential means for the meningococcus to escape immune responses to the vaccine. One clear message from the Berlin conference was that I needed to collaborate with scientists who had expertise in the chemistry of bacterial cell surface sugars to complement our expertise in genetics. If there were to be any prospect of a MenB vaccine, it was critical to find out which regions of the endotoxin molecule were accessible to antibodies.

The years from 1984 to 1990 of my time at Oxford had been a whirlwind of activities in the clinic and the laboratory as well as a great deal of overseas travel. I was desperately in need of sabbatical leave to think through future research plans with protected time to research the literature and write the necessary applications for funding. Fortuitously, a colleague from Sweden, Staffan Normark, who had been recruited to Washington University in St. Louis to head up the prestigious Department of Microbiology, had contacted

---

[f] The *International Pathogenic Neisseria Conference* held in Berlin in 1990.

me to find out if I was willing to spend time there as a visiting professor. The offer included substantial facilitating funding to cover travel, housing and other necessary expenses. My request to the General Board of Oxford University for sabbatical leave, however, went down like a lead balloon. They objected strenuously, especially when it came to who would undertake my clinical and teaching responsibilities. David Weatherall, who was virtually running the Oxford Medical School, came to my rescue with a brilliantly worded letter that protested his dissatisfaction with what he referred to as the "grudging tone" of the University's General Board — adding with well-judged cunning that "… Moxon has been advised to make firm plans to take a sabbatical." With his support I obtained some financial backing from the Wellcome Trust to support a locum clinician for one year. With the birth of Timothy in 1997, we were now a family of five. Christopher opted to remain as a boarder in Oxford until the end of the school year, but Marianne, Sarah (aged 9), Timothy (aged 3), and I took up temporary residence in the mid-western city of St. Louis.

Reference

[1] De Chiara *et al.* Genome sequencing of disease and carriage isolates of *H. influenzae* identifies discrete population structure. *PNAS*, 2014, 111:5439–5444.

Sabbatical in the Mid-West

Although I had lived on the east coast of the United States for 14 years, my sabbatical at Washington University, St. Louis, brought home to me how little I knew about the mid-west. A border State between the North and the South, populated by both Union and Confederate sympathisers, Missouri was politically divided during the American Civil war (1861–1865). The horrors of this cataclysmic war in the words of Mark Twain (1873), "... wrought so profoundly upon the entire national character that the influence cannot be measured short of two or three generations." In 1990, after five generations, the consequences were still resonant.

St. Louis boasts the largest brewery in the nation and the breathtaking Gateway Arch whose construction ranks as one of the great engineering feats of all time. The vast university campus was quite different from anything I had ever experienced and was a powerhouse of scientific activity. I will always be grateful to Staffan Normark who provided an exceptional environment for this important year. Despite his hectic agenda as Professor and Head of Microbiology, he always seemed to have time for in-depth scientific discussions. As a family, we enjoyed the internationally acclaimed Botanical Gardens, the Zoo and the numerous local city festivals and markets. I now had time to catch-up on my backlog of scientific reading and to do experiments at the laboratory bench for the first time since leaving Johns Hopkins.[a] It was also a

---

[a] I was investigating the possibility that *Hi*-b, considered to be a quintessential extracellular bacterial pathogen, might also sequester within phagocytic cells. Working with a *macrophage* cell line, I observed only very occasional instances of intracellular survival and I did not ascribe much significance to this observation until, many years later, collaborative studies with a colleague, Professor Marco Oggioni, made sense of what I had failed to understand (see Ref. 1).

joy to have time to meet with so many brilliant and engaging scientists who got me up to date with new laboratory techniques.

Lecturing in the nearby Department of Biology, I was hosted by the well-known expert on Salmonella infections, Roy Curtiss, a long-haired, bearded mid-westerner whose apparently relaxed demeanour concealed his deeply ambitious and fiercely competitive dedication to research. Wearing jeans and T-shirts, he enjoyed the status of a guru, seemingly always surrounded by admiring post-docs and graduate students whose typical relaxation involved outings on motor bikes, camping, kayaking on the Missouri with lots of beer and smoking "pot," as we called it then.

High on my agenda during this sabbatical were activities that would facilitate a switch in the focus of my Oxford laboratory. The imminent implementation of *Hi*-b conjugate vaccines meant that there would now be a compelling tide of enthusiasm to develop similar vaccines against meningococcal meningitis. With this idea firmly in mind, I went to lecture at the National Research Council (NRC) in Ottawa at the invitation of Harold Jennings, one of the world leaders in meningococcal research. A British scientist with an acerbic north-of-England bluntness, Harold had been part of the brain drain.[b] The NRC had an international reputation for carrying out detailed research on surface structures of bacteria, especially those of importance to human and animal infections. In the 1970s, Jennings was the first to investigate the structures of the several different meningococcal capsular polysaccharides.[c]

Unlike *Hi*, where only strains expressing the serotype b strains account for almost all cases of meningitis, the meningococcal strains causing meningitis express one of five distinct capsular polysaccharides, named A, B, C, W and Y, each capital letter designating a specific polysaccharide structure. Because the existing analytical techniques to define their chemical make-up were inadequate, Harold used the powerful technique of *Nuclear Magnetic Resonance (NMR)* to investigate these complex molecules. NMR involves putting a tiny amount of a substance to be investigated in a special tube that

---

[b] The term was adopted in the 1960s in the context of increasing concerns within the UK that the country was losing skilled scientific and engineering personnel to other countries, notably North America.

[c] Like the pneumococcus and *H. influenzae*, the first description of a meningococcal capsular polysaccharide was by a Rockefeller scientist (see Ref. 2).

is placed inside a hood containing a powerful magnet, similar to that used for doing MRI scans in hospitals.[d]

Harold Jennings' pioneering research on the meningococcal polysaccharides was brilliantly successful and elucidated for the first time the structures of the three capsular variants (A, B and C) most often responsible for causing meningitis. His commitment to research on meningococcal meningitis had been sparked by several devastating outbreaks of meningococcal C (MenC) meningitis in Canada. Harold had discussed the problem with Emil Gotschlich — the pioneer of meningococcal polysaccharide vaccines that prevented meningitis in the military. Since these vaccines did not protect young children efficiently, Gotschlich encouraged Harold to develop meningococcal conjugates that, based on the success of *Hi*-b vaccines, would likely protect infants.

By 1981, Jennings had conjugated three of them (the A, B and C meningococcal capsular polysaccharides) to tetanus toxoid, a carrier protein chosen because it had an exemplary safety record as a vaccine routinely given in national immunisation programmes. The MenA and MenC capsular polysaccharide–tetanus toxoid conjugates induced excellent protection in animal models. In stark contrast, the conjugated MenB polysaccharide failed to induce antibodies, a problem that had been worrying scientists for years: why was this polysaccharide such a poor immunogen?[3]

Crucial insights[e] had come from scientists in Finland who noted that the chemical structure of the MenB capsular polysaccharide was identical to cell

---

[d] The nuclei of the atoms of key elements (e.g. carbon, hydrogen or oxygen) are surrounded by orbiting electrons, charged particles which generate a small magnetic field that partially shields the nuclei from the much more powerful external magnetic field created by the NMR machine. The amount of shielding varies according to the exact properties of the components of the material being examined. For example, hydrogen bonded to oxygen will be different from hydrogen bonded to carbon and these differences generate characteristic spectra, a distinctive "fingerprint" that allows the chemist to determine the precise structure of the substance — in this instance a polysaccharide.

[e] Taking advantage of the magnificent John M Olin Library of Washington University, I learned that the B polysaccharide was a polymer made up of repeating units of sialic acid and therefore called poly-sialic acid or PSA. Present on the surface of human cells. PSA modulates cell to cell communication resulting in profound effects on the development of the mammalian nervous system. Antibodies or enzymes that modify PSA affect the migration of nerve cells, neuronal connectivity and the formation of junctions between muscle and nerves. Therefore, inducing antibodies to this self-antigen posed a potentially unacceptable safety risk.

surface molecules (*glycoproteins*) of human cells, especially those found in the brain. They proposed that because of this mimicry, immune responses to the B polysaccharide were subject to an immunological taboo, called tolerance, a restraining order to prevent the immune system from attacking the body's own tissues. If the Finnish scientists' ideas were correct, the implications were far-reaching. A vaccine that overcame tolerance could induce antibodies that were dangerous and implied the need for a completely different approach to the development of a MenB vaccine.

Jennings saw this as a scientific challenge and thought that chemistry could provide a means to solve the problem. His plan was to make a derivative of the B polysaccharide that would avoid the potential damaging cross-reactivity to human cells and permit an effective antibody response. John Robbins was strongly supportive of this approach, but many leading internationally respected scientists were sceptical. In the end, it was the vaccine manufacturers' opinion that was the showstopper. They anticipated that ethical concerns over safety would block the possibility of doing clinical trials on any vaccine based on the B polysaccharide.

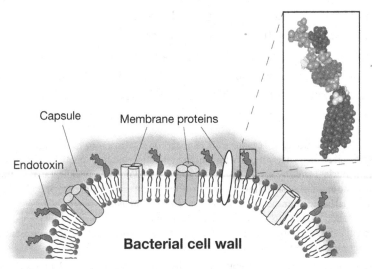

Figure 15.1   A more detailed diagram of the endotoxin molecule. The lower portion that is closest (proximal) to the cell wall is often referred to as Lipid A. It is inserted into the outer membrane of the bacterial cell. Composed mostly of lipids (fats), this is the region of endotoxin that is responsible for tissue injury through inciting inflammation when it comes in contact with human tissues. The outer or more distal regions of endotoxin are composed largely of sugars. These are potential targets for antibodies and therefore a potential basis of a vaccine.

An alternative idea was to explore the candidacy of meningococcal endotoxin (see Figure 15.1) as a vaccine, an approach that I hoped would be of interest to Harold Jennings. Endotoxin is a complex molecule consisting of lipids (fats) and sugars. The lipid portion anchors the molecule into the bacterial membrane. This component is extremely damaging to human tissues (hence the name, endotoxin) and is one of the major factors resulting in death from meningococcal sepsis and meningitis. At the peak of my patient Julia's illness (see Chapter 1), the millions of bacteria in every teaspoonful of her blood resulted in high concentrations of endotoxin. The potentially injurious lipid component was not what I had in mind as a vaccine! Rather, I was interested in the sugar structures that project outwards from the bacterial cell made of sugars, such as glucose, galactose and sialic acid. I hoped that through collaboration with Harold Jennings, NMR could determine how these sugars were spatially arranged on the surface of the bacterium, crucial information required to develop a vaccine. But Harold did not have the time to work on this project and suggested that I collaborate with one of his senior colleagues. This turned out to be a great decision; Jim Richards and I immediately took a liking to one another and began a collaboration that would endure for almost two decades. I'll return to this project later.

Another idea for a MenB vaccine was targeting proteins on the bacterial cell surface. At magnifications 200 times greater than conventional light microscopy, electron microscopy showed that meningococci made spherical blebs that were shed intact from the bacterial surface (Figure 15.2). The idea was to use these membrane blebs as a vaccine. The tissue damaging lipids mentioned above could easily be removed by treatment with a detergent to provide detoxified outer membrane vesicles (OMVs). When injected into animals, OMVs induced antibodies that were protective. This was encouraging although there was a major snag. Because of the immense variability of these surface proteins, the OMVs only protected against a limited number of meningococci. One of the pioneers of the OMV vaccine, Jan

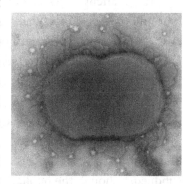

Figure 15.2   Electron micrograph of a meningococcus showing shedding of outer membrane blebs.

Credit: Photograph kindly provided by the Public Health Laboratory Service Centre for Applied Microbiology and Research, Porton UK.

Poolman, made a panel of antibodies to index variations in the membrane proteins of MenB strains that had caused meningitis in Holland since the early 1980s. He thought that by selecting a mixture of OMVs from about six different kinds of MenB, he could make a successful vaccine.

Lithe and boyish in appearance, Jan was the lead scientist at the Netherlands National Institute for Public Health. I had met him at the international meeting in Berlin (just before leaving for my sabbatical). We had discussed the MenB epidemic in Norway where, at the time, one case of meningitis per year occurred in every two thousand people. In one household alone, three out of four children had died from meningococcal meningitis. The government's advice to parents was that any baby that went to bed with a fever should be woken through the night to facilitate rapid detection of the dreaded MenB infection. Households, especially those with young children, were terrified.

There was an urgent need for a vaccine and Jan was working with Norwegian and US scientists who were pinning their hopes on an OMV vaccine. But after months of work, the results were disappointing. As predicted, the variations in the target meningococcal protein (called PorA) were a major stumbling block. But then came a crucial finding that was a turning point. Almost all the isolates from the Norwegian epidemic were the same genotype and shared an identical PorA protein; an effective OMV vaccine was in fact entirely feasible.

The challenge was making enough OMV vaccine for a population of four million. The Norwegian Institute of Public Health's sole manufacturing unit was one small, ludicrously inadequate basement room. So, the Institute Director courageously invested his entire budget, expanded the facility and full-scale trials of the vaccine began. From 1989 to 1991, 1500 Norwegian schools participated, suspense mounting as the case files were sorted into two piles, those who had received the vaccine and those who had received a placebo. The vaccine efficacy was 57%, hardly the triumph that had been expected.[f] The Institute Director's television announcement to the Norwegian public was subdued, although the public's reaction was positive.

Norway was not alone in suffering from epidemic meningococcal disease. Cuba too had prepared an OMV vaccine to combat heightened rates of MenB

---

[f] Preliminary results from a small sample taken at 10 months had been more encouraging, but the definitive results were less impressive than had been predicted.

meningitis and by early 1990 had seen a marked decrease in cases of the disease. Their success was duplicated in Brazil and Chile. But these clonal MenB outbreaks, amenable to intervention using OMV vaccines, were very much the exception. In most countries, MenB meningitis was caused by many distinct strains for which the OMV vaccine lacked the required broad coverage. The need for a "universal" meningococcal vaccine was evident and became one of the highest priorities of the vaccine research and development portfolios in academia and industry.

References

[1] Ercoli, G. *et al.* Intracellular replication of *Streptococcus pneumoniae* inside splenic macrophages serves as a reservoir for septicaemia. *Nat. Microbiol.*, 2018, 3:600–610.

[2] Rake, G. Studies on meningococcus infection: 1. Biological properties of fresh and stock strains of the meningococcus. *Journal of Experimental Medicine*, 1933, 57:549–559.

[3] Wyle, F.A., Artenstein, M.S., Brandt, E.C. *et al.* Immunologic Response of Man to Group B Meningococcal Polysaccharide Vaccines. *Journal of Infectious Diseases*, 1972, 126:514–522.

The Conjugate Vaccines and the Formation of the Oxford Vaccine Group

My sabbatical over, I returned in the summer of 1991 to oversee the trial in which the *Hi*-b conjugate vaccine was offered to all children[a] in four of eight regions of Oxfordshire. Its outcome was spectacular: there were no *Hi*-b infections among more than 10,000 immunised children compared to 11 cases in a similar number who had not been given the vaccine. Meantime, the UK Department of Health had put in place a stunningly successful education campaign to prepare for its nationwide introduction. David Salisbury had done his homework, adopting sophisticated marketing strategies. One of the most memorable publicity-clips aired on national television featured an animation of a toddler putting a doll into a coffin. Through these powerful images, parents learned that *Hi*-b meningitis could be fatal — but importantly, there was a vaccine to prevent this deadly infection.

Beginning in October 1992, *Hi*-b conjugate vaccine was offered free to all children starting at age 2 months.[1] Consistent with its impact in North America and other European countries, the conjugate vaccine resulted in a precipitate decline of *Hi*-b disease. What had been the commonest cause of meningitis in children was virtually eliminated within months (Figure 16.1). It is one of the most impressive examples of how quickly and safely a vaccine can eradicate a killer disease.

Of course, there was a need to evaluate the longer-term impact of the introduction of the *Hi*-b vaccine. My Oxford clinical trials research team

---

[a] Babies born after 1 July 1991.

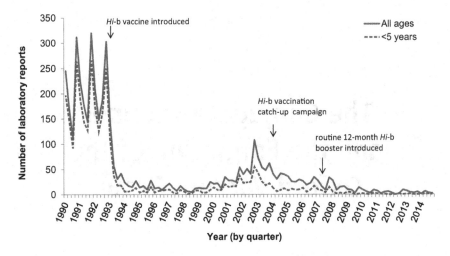

Figure 16.1 Graph showing *Haemophilus influenzae* type b laboratory reports by quarter: England, 1990–2004. Note that only children aged up to 5 years (dashes) were immunised (direct protection). The commensurate precipitate decrease in cases at all ages (solid) occurred through the reduced exposure to *H. influenzae* type b because of a reduction in person-to-person spread (indirect protection), as discussed on page 144.

set up a surveillance study to monitor its effectiveness[b] and to identify any *vaccine failures*. A number of vaccine failures were observed commencing around 2000, in part because the vaccine reduced *Hi*-b carriage and, as a consequence, there was a decrease in natural boosting of serum antibodies. The problem was eliminated by a vaccine catch-up campaign and the introduction of a booster dose of *Hi*-b vaccine in 2007 (Figure 16.1). It highlights the importance of post-implementation surveillance. There was also concern that other causes, such as immunodeficiency, might explain these failures.[c] The snag was that traditional sources of funding offered scant opportunities to finance this kind of research.

Fortunately, I knew the senior scientist of Pasteur Merieux Connaught (PMC), Stanley Plotkin, with whom I had strong connections dating back to my time in the US. A polymath, as erudite discussing Shakespeare or philosophy as he was skilled in laboratory research on viruses, Stan was ten years my senior and had been the Head of the Paediatric Infectious Diseases Department at the

---

[b] After the introduction of the vaccine (October 1992), more than 600 cases of *Hi*-b meningitis were prevented each year in England and Wales.

[c] This initiative, a collaboration with the Department of Health, was led by a succession of trainees (Robert Booy, Paul Heath and Jim Buttery).

Children's Hospital of Philadelphia. A stellar name in the field of vaccines, he had been the pioneer of the vaccine against German measles that had prevented crippling birth defects (congenital rubella) in millions of children as the result of exposure of their mothers to the virus during pregnancy.[2]

PMC was making millions from sales of their *Hi*-b vaccine that had been greatly facilitated by our clinical trials. My pitch to Stan was that PMC had benefitted from our research capability in Oxford and could do so in the future. But for there to be a thriving university-based research group doing vaccine trials and associated research, I needed prospective funding to maintain the essential personnel and operating costs of my unit. Could we do a deal whereby PMC would commit up-front funding while, in turn, we would provide them with a first option to do clinical trials of their vaccines? In truth, I felt a bit as though I was going to one of the major banking institutions to ask for an interest free loan on the grounds that I represented a good investment.

There were obvious weaknesses and risks to my proposal — for both PMC and Oxford. But by the 1990s, the ambience for forging links between academia and industry was becoming a much more viable pathway than it had been a decade or so earlier. The idea that a university should be constrained by its traditional commitments to teaching and blue-skies research, while leaving applied research to the commercial sector, was a fast receding concept. In my own university, two academics in biochemistry and neuroscience had forged long-term lucrative (multi-million pound) commercial contracts with "Big Pharma."

At the time (early 1990s), Prime Minister Thatcher had made clear her commitment to funding basic research, but she was critical of the failure of British scientists to capitalise on the commercial revenues of scientific discoveries. A good example was the discovery in 1977 of *monoclonal antibodies* for which Cesar Milstein of Cambridge University later received a Nobel Prize. Because of a lack of expertise in technology transfer within academia, muddled treasury thinking and bureaucratic delays within government, the technology was not patented. Not only did the government refuse to provide the money to file a patent, they also refused Milstein to do it himself on the grounds that he had received funding from the Medical Research Council (MRC) and did not own the intellectual property. But a rival American scientist did file a patent and it generated billions of dollars that benefitted hard-nosed opportunistic entrepreneurs in the US.

However, there was still resistance within academia for "getting into bed" with commercial companies — seen by some as sullying research independence. The counterargument was that without thriving commercial–academic partnerships, much essential research would simply not get done. Universities, forever short of money, were waking up to the huge potential of commercial funded research collaborations that large pharmaceutical companies were keen to foster.

Nonetheless, my idea to approach industry to fund my research in the 1990s incited concerns. These included whether an academic research group receiving commercial funding could steer clear of conflicts of interest and how ethical committees reviewing clinical trial protocols would react to research collaborations of this kind. Who would "own" the research data and what sort of intellectual property agreement would be appropriate and acceptable? Would a contract oblige us to carry out a clinical trial even if there were disagreements on the scientific basis of the study design? On the PMC side, there were concerns that any research agreement, acceptable to Oxford University, would offer no clear-cut advantages over their current *ad hoc* activities. My counter was that PMC's clinical trials done by Oxford would be of higher quality and be completed more rapidly.

Stan Plotkin and I met to discuss how we could forge a collaboration based on two research themes. The first centred on the completely unanticipated finding[d] that the *Hi*-b conjugate vaccines not only protected immunised individuals from disease (direct protection), but also reduced the spread of *Hi*-b organisms. This was a major surprise and hugely important. The rationale underpinning the *Hi*-b conjugates was to induce antibodies that killed or helped remove bacteria from the blood stream thereby preventing their dissemination to the brain. Antibodies in the blood were not expected to have any effect on transmission of *Hi*-b bacteria from person to person via secretions (mucus, saliva, etc.). But they did, showing that the vaccine conferred not just direct protection of those immunised but also indirect protection (*herd* or *community immunity*) through reducing spread. But this bonus also introduced a potential downside. Reducing colonisation of the nose and throat with *Hi*-b resulted in a loss of natural boosting of antibodies, as discussed above, but

---

[d] Several independent studies from the US, Finland and UK.

also there was the potential for other bacteria to take over the vacant niche perhaps with virulent non-b capsular variants of *Hi*. It was a reminder of the importance of the Professor's question at my inaugural talk at Johns Hopkins.[e] I was pleased when Stan Plotkin agreed on the importance of continuing the national *Hi*-b surveillance.

A second important topic of potential collaboration was the need for clinical trials of other PMC vaccines. These were costly and, in Stan's opinion, were often done poorly within the industry. The bottom line was that he was willing and enthusiastic to work together on a collaboration between Oxford and PMC. The plan included regular meetings and brainstorming sessions, a platform for mutual exchange of scientific ideas. I was training future infectious disease paediatricians who would benefit greatly from these sorts of interactions. From Stan's point of view, many of his industry scientists were medical doctors who lacked any specialised knowledge in vaccine research. Mutual needs could be met, although there was a caveat. Vaccine manufacturers were subject to very tight timelines, imperatives that might not go down well with academia. Had I oversold my hand? Could we deliver what was needed and how would the University react?

Another thorny problem was how to dove-tail our research activities with those of the UK Department of Health, who were responsible for drawing up national vaccine policies. Policies are shaped by evidence, so there was inevitable overlap between the research activities in Oxford and those of the Department of Health. The purchase of vaccines for the UK — the *Hi*-b conjugates had been a case in point — was the responsibility of the Department of Health who had to select and negotiate the best deal in terms of both quality and price, a necessarily sensitive and highly confidential process. This placed constraints on transparency and open discussions, not at all a healthy situation from the perspective of academic researchers. I decided to discuss my plans for a clinical trials unit with two key colleagues[f] in the UK Centre for Disease Control in Colindale. Their main concern was how I planned to avoid conflicts of interest. I had already discussed this at length with Stan Plotkin and had made it clear to him that any collaboration with PMC could not be exclusive

---

[e] See Chapter 8, p. 80.

[f] Norman Begg and Elizabeth Miller, senior scientists who had a wealth of experience in infectious diseases epidemiology and vaccines.

and that we would also be working with other vaccine manufacturers. This, I argued, was in the best interests of PMC as it would counter criticisms that we were in the pocket of any one vaccine manufacturer.

Predictably, an account of my Colindale meeting filtered through to the Department of Health's Immunisation Division who were not at all happy about my idea to set up a vaccine trials unit in Oxford, part-funded by the commercial sector. But I could see few other options for getting funding. I now had to build-up our reputation for vaccine research, and as part of the strategy a name for the vaccine trials unit was needed. I came up with the *Oxford Vaccine Group (OVG)* and in 1993 I submitted a formal proposal to the University's General Board for it to be recognised as an independent research unit within the Department of Paediatrics. It soon became clear that I had underestimated the maelstrom of problems that I would encounter. Oxford University has its own frustratingly slow processes and acceptance of OVG as a legitimate entity took prolonged negotiations. After months of haggling, a deal was struck enabling the OVG–PMC collaboration to go ahead. The formal agreement centred on careful post-implementation surveillance of the *Hi*-b conjugate vaccines and clinical trials of some new vaccines that included a combination diphtheria, pertussis, tetanus (DPT) and *Hi*-b vaccine. This formulation reduced by half the number of "jabs" given to infants in the early months of life.

The escalation of meningococcal disease in the UK as well as other European countries brought other exciting opportunities for OVG. In the early 1990s, outbreaks of a highly virulent strain of meningococcus expressing the C capsular polysaccharide were occurring in the UK, France, Greece, Spain and, across the Atlantic, in Canada. The work of Harold Jennings and several other research groups had led to testing meningococcal conjugates in the laboratory, but there had been no human studies. The crucial issue was how meningococcal conjugate vaccines could be taken forward into human trials without a formal commitment by governments for vaccine purchase. Vaccine manufacturers were struggling to maintain commercial viability and needed reassurance that they would recoup research and development costs. In response, the UK Department of Health had formed and funded a National Vaccine Evaluation Consortium whose objectives were to generate the data needed to support the licensure of vaccines and to inform policy decisions about their potential use. The MenC conjugate vaccines were a major focus of the consortium and provided the required incentive for the vaccine manufacturers; if their vaccine passed muster, it would be purchased for use in the UK routine immunisation programme.[8]

I set up a collaboration for OVG to do a major clinical trial of Chiron's MenC conjugate vaccine. After obtaining the agreement of NHS General Practitioners and local ethical approval, we contacted the parents of more than a thousand children born in Oxfordshire and enrolled 182 infants, some 14% of those approached. The main reason for parent's refusing to be part of the trial was the need to obtain four blood samples from each child, a real challenge of trust and commitment given that these babies were only a few months old. About half a teaspoonful of blood on each occasion was needed to carry out the necessary laboratory tests. There were understandable parental concerns. It takes a lot of courage and considerable *altruism* for parents to allow their recently born babies to receive a new and experimental vaccine and so many blood tests. Recruitment involved a huge number of telephone calls and sometimes preliminary visits to the home to answer questions and to establish *rapport*. Not unexpectedly, there were sometimes family tensions where the mother and father were not in agreement about their child's involvement. The mother usually had the final say!

An unusual aspect of the Oxford Vaccine Group's approach was making home visits to give the vaccines, take blood samples and obtain the patient data required to comply with the regulatory procedures for licensure. Based on extensive previous research, we had analysed the feasibility of home as compared to clinic visits and the results were crystal clear; recruitment and compliance were hugely greater if we opted to go to the home. Indeed, many of the parents used the visits to talk about their children's health and seek advice about immunisations and illnesses for siblings that were not involved in the trial. It was crucial to ensure that parents were well-prepared for the visits. For example, getting their cooperation in applying the anaesthetic cream to the babies' skin well before arrival of the trial team to facilitate taking blood without wasting time. In obtaining blood samples, it was a firm rule that only one attempt would be made, putting a lot of pressure on the nurses and doctors

---

[8] Three vaccine manufacturers became involved: Baxter Biosciences who had purchased the rights to make the conjugates developed in the Jennings' laboratory, and Wyeth (formerly American Cyanamid) and Novartis who had acquired the Chiron Vaccine conjugates developed by Paolo Costantino in Siena.

involved, although success rates were better than 90%. A missed sample meant crucial absence of data. Successful interactions with families were helped by keeping careful records. Did the visit go well? Were the baby, parents or siblings upset? It was helpful to remember birthdays, likes and dislikes — details that encouraged good interactions.

Home visits had advantages, but were also sometimes challenging; for example, coping with the distractions and disruptions of siblings. The condition of houses varied with respect to cleanliness, the presence of many rats on one occasion being one of the less welcome experiences for the OVG researchers. One family was housed in a canal boat that could only be accessed by crossing a field after a substantial snow fall. The team had to be prepared for the unexpected, including strange pets and one 'rough diamond' husband who did not like the idea of vaccines and threatened to fetch his shotgun if the team did not leave immediately. Indeed, training in conflict management was an important component for OVG researchers. For example, making sure that their car was not blocked in and that mobile phones were pre-set to send out an SOS. Travel to the homes required huge preparation of itineraries to maximise the number of visits. One day, a traffic jam blocked the M40 motorway for several hours and all the planned visits had to be cancelled. Some inevitable frustrations occurred, such as arriving to find the household were out or away on holiday, although every effort was made to contact the parents by telephone prior to the planned visit. Mishaps inevitably occurred: a collision with a deer, a flat tire, a speeding ticket, a burst radiator on a hot day and running out of petrol due to a faulty fuel gauge. But most of the time, the visits were uneventful and, according to a questionnaire completed by GPs, parents enjoyed participating and contributing to the research.

MenC conjugate vaccines were safe and induced antibody responses in children aged as young as 2 months. The Chiron trial, especially complex in its design, was published in the *Journal of the American Medical Association* and an accompanying editorial commented on the robust conclusions based on two features of the design. The use of a functional laboratory test showing that the vaccine-induced antibodies were protective and that, several months after the immunisations, the children responded by making new antibodies when they were given a booster dose. This indicated that the vaccine had induced immunological 'memory' resulting in long-lived protective immunity.

Based on these sorts of studies, most of which were undertaken by the National Vaccine Evaluation Consortium, the regulatory authority (European Medicine Agency) agreed to license several MenC conjugate vaccines. This was an unprecedented decision as conventionally the pathway to licensure includes at least one clinical trial to show that a vaccine prevents disease in people. But efficacy studies are hugely expensive, require very large numbers of children and often take years to complete. Given that the correlation between laboratory assays and protection had been so rigorously established, it was argued that these data were an adequate basis for licensure. By September 1999, the major hurdle for their national implementation in the UK was cleared.

But solid science was not all that was needed; political will was also crucial and the individual who provided this was Frank Dobson, "… of portly frame, jovial expression, and bright white beard,"[h] appointed as Secretary of State for Health by Tony Blair following Labour's landslide victory at the 1997 general election. Although this was a high-profile post, Dobson was frustrated by interference from civil servants and wrote a memo to Blair, saying "If you want a first-class service, you have to pay a first-class fare — and we're not doing it." In 1999, he was confronted by one of his constituents whose child had died from meningococcal meningitis. The angry parents told Dobson that the tragedy could have been prevented by a vaccine. Dobson called David Salisbury in the Department of Health's Immunisation Division: both saw an opportunity to make a major contribution to public health, one in which they would play a crucial role on the international stage; the UK could be the first country to implement MenC conjugate vaccines. But how was the money to be found given the difficulties in loosening the Treasury purse strings? To raise extra money from the Health Department's resources, Dobson froze all departmental spending until the required money was found from existing budgets. But just when all the pieces of the complex plan seemed to be in place, it was challenged on the grounds that any health intervention needed to be justified on economic grounds — using the so-called cost-effectiveness analyses demanded by the *National Institute for Clinical Excellence (NICE)*. "Sod the economists," Dobson barked in his characteristically pugnacious style.[i]

---

[h] Frank Dobson won the *Beard of the Year Award* in 2000.

[i] In fact, no published cost-effectiveness study was carried out prior to national implementation of the vaccine, although data were published post-implementation. More than a decade

There was then another problem; a leading manufacturer couldn't produce enough of the MenC conjugate vaccine to begin the programme on time. Dobson argued for more vaccine, but Wyeth told him that they couldn't provide it. David Salisbury refused to take "no" for an answer and, fearing the consequences of adverse publicity, Wyeth agreed to increase their output by 75%. By October 2000, the UK programme was fully launched with spectacular results. Within a few months, cases of MenC meningitis decreased by close to 100%. The Department of Health scientists published their results in what was a milestone in vaccinology, a truly remarkable achievement,[3] especially since the vaccine had been implemented without the traditional large-scale efficacy trial on which some prominent scientists had insisted. Further, there was a huge additional benefit; as had been found with *Hi*-b conjugates, the vaccine also sharply decreased person-to-person spread. This indirect community protection (herd immunity) accounted for half of the protection against MenC meningitis.

However, MenB was still by far the dominant cause of deaths and disabilities from meningitis in the UK and many other countries. Jan Poolman's efforts to develop a multivalent MenB vaccine incorporating several PorA antigens had run into problems, and in 1997, frustrated by the inadequacy of Dutch government funding, he left the Netherland's National Institute for Public Health (RIVM) to become Head of Bacterial Vaccines at Smith Kline Beecham Biologics[j] in Rixensart, Belgium. The need for a universal MenB vaccine was high on the public health agenda and he hoped GSK would give him adequate support.

In the present chapter, I have summarised the activities of the Oxford Vaccine Group and the impact of *Hi*-b and meningococcal conjugate vaccines during the 1990s. In the next chapter, I must go back several years to pick up on what had been happening in my laboratory after returning from my sabbatical in 1991. The MRC had awarded me a programme grant to investigate the potential of *Hi* and meningococcal endotoxin for developing future vaccines.

---

later, cost-effectiveness analyses were a major factor in the decision of the Joint Committee on Vaccines and Immunisation (JCVI) on whether to implement a MenB vaccine.

[j] Shortly to become GlaxoSmithKline (GSK).

References

[1] Heath, P.T. and McVernon, J. The UK vaccine experience. *Archives of Disease in Childhood*, 2002, 86:396–399.

[2] Wadman, M. *The Vaccine Race. How Scientists Used Human Cells to Combat Killer Viruses.* Penguin Random House, 2017.

[3] Campbell, H., Borrow, R., Salisbury, D., and Miller, E.S. Meningococcal C conjugate vaccine. The experience in England and Wales. *Vaccine (supplement)*, 2009, 27:B20–B29.

References

The New Genetics and Genome Sequencing

Centuries ago, folklore and experiences such as the Plague of Athens taught us that once a person has been infected by a microbe, they subsequently resist it much more efficiently through what is called immunity. Vaccines induce immunity by deliberately exposing a person to a harmless form of the germ (or a fragment of it). Although many excellent vaccines are based on weakened (attenuated) versions of the whole infectious organism (for example the viruses of smallpox, measles and polio), there is merit in narrowing down the components to just one, or a very few components (antigens), of a pathogen. Logically, it makes sense that many of the most effective vaccines are based on virulence factors, the microbial components that are directly involved in the disease process, for example, bacterial toxins or the capsular polysaccharides that have been so central to this story. Molecular biology, including methods of cutting (cloning) and pasting (recombining) DNA, provides a means to identify and modify virulence genes (see Chapter 10) that code for proteins that make highly effective vaccines.

As already discussed in the previous chapters, a major component of the outer membrane of many bacterial cells is endotoxin (see Figure 17.1), a complex molecule that is involved in the implantation (colonisation) of *Hi* and meningococci in the nose and throat, invasion of the blood and dissemination of the bacteria to the meninges. Endotoxin is also a major factor in causing tissue damage through inciting inflammation. To make headway on investigating the potential of endotoxin as a vaccine, a genetic approach was the best way to identify the components that would induce protection while avoiding those, especially the endotoxin lipids embedded in the membrane, that are injurious to body tissues.

Based on theoretical estimates, the genomes of *Hi* and meningococcus each had a total of around 2,000 genes. About 40 of these genes were required to produce and assemble the different lipid and sugar components of the endotoxin molecule on the bacterial cell surface. At the time, very few of these genes had been identified using classical genetics. The existing methods were time-consuming and usually failed. Instead of banging our heads against a brick wall, my lead scientist (Derek Hood) and I realised that the logical way forward was to obtain complete genome sequences of the bacteria. It would give us inventories, a sort of "yellow pages" directory, that included all the genes required to make the endotoxin molecule.

However, in the early 1990s, the idea of sequencing the entire genome of a bacterium was an unrealistic pipe dream. In the late 1970s, the pioneering methods of the Nobel Prize laureate Frederic Sanger had sequenced some viruses of around five thousand nucleotides (hundreds of times smaller than a bacterial genome). His technology resulted in the first generation of commercially available sequencing machines that by 1990 were becoming widely available[1]

Endotoxin molecule

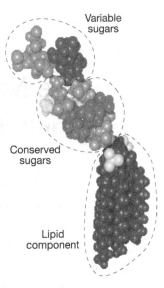

Figure 17.1   A molecular model of the endotoxin molecule of meningococcus B. It can be considered to have three regions. The lipid component is inserted into the cell membrane. Projecting outwards from the cell wall are the conserved sugars that were considered as a potential vaccine antigen. The outermost sugar components were too variable to be a suitable target for making a vaccine. The author wishes to acknowledge J.R. Brisson and Jim Richards of the Department of Biology, National Research Council, Ottawa, Canada for generating the molecular model of endotoxin.

in research laboratories. US scientists had started to sequence the bacterial genome of *Escherichia coli*,[a] but completion of the project was years away. A consortium of several laboratories was sequencing the genome of an important

---

[a] The lead scientist was Fred Blattner, Professor of Genetics at the University of Wisconsin. The choice of *E. coli* was because this was widely accepted as the model microbe as biologists knew more about its physiology and biochemistry than any other cell. Many Nobel Prizes have

soil bacterium (*Bacillus subtilis*). Since the genomes of meningitis bacteria were less than half the size of *E. coli* or *B. subtilis,* much less work would be needed to sequence them, so I considered organising a consortium of research laboratories to sequence the *Hi* genome.

Who better to discuss this with than my former mentor at Johns Hopkins, Hamilton ("Ham") Smith. An invitation to lecture at Johns Hopkins in Baltimore provided a perfect opportunity, and in July 1993 I returned to the laboratory in which 15 years earlier I had started my work on the *H. influenzae* b (*Hi*) capsule genes. I felt quite emotional as I entered the familiar surroundings where my career as a scientist had been so dramatically changed in the late 1970s. Keats's poem (*On First Looking into Chapman's Homer*) came to mind.[b] Recombinant DNA technology had completely changed my vision of what could be achieved in the biomedical sciences, specifically in research on bacterial pathogens. Now, in 1993, here I was once again face to face with Ham to talk to him about the genome sequence of *Hi* about which I had sent him a draft proposal as a basis for discussion.

I was totally unprepared for what happened. Ham was quietly but firmly dismissive of my consortium idea, arguing that I had completely underestimated the amount of work that would be involved. Anyway, he was adamant that it should be done in dedicated DNA sequencing facilities, not by small research laboratories funded to do basic science. I felt disappointment and embarrassment in equal measure. Ham was by nature rarely dogmatic or aggressively critical, but on this he was as scathing as I had ever known him. Further, he was not forthcoming when I pressed him to tell me where he thought the large-sale sequencing could be done. Later, I came to appreciate that Ham was going through something of a personal crisis. In the aftermath of his Nobel Prize, he had struggled to live up to his sudden celebrity. He found it difficult to know what research directions to take after his seminal discovery of restriction enzymes. This dark period was further exacerbated by preoccupation with his brother who had been diagnosed with schizophrenia and with whom he was very close. The possibility that his sibling's illness was

---

been based on research on *E. coli*, including understanding the genetic code, mechanisms of replication, gene organisation, regulation and the basis of mutations.

[b] "Then felt I like some watcher of the skies/When a new planet swims into his ken."

genetically determined and that he too might be susceptible was the cause of huge anxiety.

It turned out that Ham had not shared something important with me over the *Hi* genome project idea because, as he told me later, its potential was at the time of our 1993 July meeting far too speculative. Three months earlier Ham had met Craig Venter with whom he had struck up a friendship. They were a study in contrasts; Ham, shy and self-effacing with quiet penetrating intelligence, was intrigued by the self-confident, brash personality of Venter — destined to become one of the most influential scientists of the late 1990s and early years of the next century.[c] As a young man, rather than attend college, Craig had spent his days hanging out on Newport Beach in California where he became hooked on surfing until in 1967, aged 21, he was drafted and served as a Combat Medic in the US Navy. In Vietnam, he experienced at first hand the appalling experiences of war and the horrors suffered by soldiers and civilians who were so badly injured that they hardly knew if they were dead or alive. He talked with amputees who confided in him that they would rather have died. As Venter himself recounts in his biography,[2] lonely and disillusioned, he swam far out to sea off the China Beach, momentarily with the intention of ending his life. Suddenly, consumed with fear and fuelled by adrenalin, he swam back to the shore.

On returning to the US, profoundly affected by his experiences in Vietnam, Craig enrolled in medical school at the University of California, San Diego, but soon switched to basic science, obtaining a degree in Biochemistry and a PhD in Physiology and Pharmacology. By the late 1980s, he had become one of the leading scientists at the National Institutes of Health (NIH) located just outside Washington DC. Charismatic and egocentric, he had no problem trumpeting his ideas and achievements or disguising his desire to win a Nobel Prize. He made many enemies, lived extravagantly and incurred the disapprobation of those who likely envied his success. But his maverick personality, his scorn for those unwilling to challenge accepted wisdom, his energy and intellectual acuity won him many admirers.

Ham and Craig had met for the first time at a scientific meeting in Bilbao, Spain, in March 1993, a few months before my anti-climactic meeting with

---

[c] The Human Genome Project was completed in 2003.

Ham. In the bar at the end of the day's scientific sessions, Ham jokingly asked Craig, "Where are your horns, because in academia you're considered to be a devil." Craig was, at the time, at the heart of a bitter controversy about his highly original approach that had enabled him to identify more than 10% of all known human genes in just a few months.[d] It was a phenomenal achievement that sent shock waves through the scientific community. The genome czar, Jim Watson, was one of those who was badly shaken. Watson felt threatened because Craig's research provided a short-cut to knowing how many genes there were in the human genome, one of the key rationalisations underpinning the Human Genome Project.

Craig's research had sparked a major row within NIH over whether these genes could be patented and who would benefit from the intellectual property. The egos of the major NIH decision-makers, the bickering over the ethics and profiteering of human gene patents proved too much for Craig. He was finding the tensions between academic freedom and commercial imperatives highly problematical and decided to set his sights on a different vision of how he would do his research. In a characteristically audacious and risky decision, he left the National Institutes of Health and set up The Institute for Genome Research (TIGR),[e,3] a not-for-profit company, financed by venture capital.[f] It was a Faustian deal that would later cause him (and, as it turned out, me) significant problems that I describe later.[g]

Over a dinner in Bilbao, Craig had invited Ham to join the Scientific Advisory Board (SAB) of TIGR. Ham was intrigued by what Craig had told him about his sequencing facility, although at the time of the July meeting

---

[d] The research had been published in 1991. The key principle was to isolate the messenger RNA from different kinds of cells and from its sequence deduce the corresponding DNA. Known as *expressed sequence tags (ESTs)*, the beauty of his approach was that although every cell in the body has an entire copy of the human genome (23 pairs of chromosomes), the only genes that are expressed (i.e., make proteins) are those that determine the unique functions of each of these specialised organs — brain, liver, kidney, heart, etc. Messenger RNA is the intermediate molecule through which DNA instructs the cell to make particular proteins.

[e] Siddhartha Mukherjee recounts how Venter had at first named his new Institute *IGOR*. He changed the name because of its unfortunate association with a "cross-eyed butler apprenticed to Frankenstein" (see Ref. 3).

[f] Substantial funding was provided by Human Genome Sciences whose CEO was William Haseltine, a notoriously ambitious and successful profiteer of genome sequencing.

[g] See Chapter 20.

with me, he knew little about it. All this was to change at the first SAB meeting at a river retreat on the Chesapeake Bay just south of Annapolis, Maryland, in the autumn of 1993. If Ham's expectations were initially low-key, seeing TIGR's facilities at first hand was transformative. Ham was astonished at the scale of TIGR's output: 30 DNA sequencing machines cranking out 400,000 nucleotides a day. The significance of what was happening hit him like a thunderbolt: the complete genome sequence of *Hi*, two million nucleotides, should only take a few months. When Craig Venter asked his scientific advisors to make suggestions about future projects, Ham raised his hand, "You call yourselves the Institute for Genome Research. How about sequencing a bacterial genome?"[h] Craig liked the idea and was also very keen to work with Ham whom he had liked from the outset. Besides, Ham was a Nobel Prize laureate, a great asset to TIGR's image.

Progress was slow to begin with as Ham had difficulty persuading his own group to divert time from their projects to create an ordered library of DNA *Hi* genomic fragments, the accepted first step in sequencing a genome.[i] Frustrated, he began to question whether a physical map was essential and turned his attention to a computational approach. He had been extremely impressed with TIGR's sophisticated computer software that had been developed to assemble raw sequence data. Ham had started his research career working on viruses and had not forgotten Fred Sanger's approach to DNA sequencing. Called the "shotgun" technique, Sanger had not bothered with the first step of constructing a physical map.[j] He had just sheared the viral genome into fragments and sorted the sequenced pieces as one might throw the pieces from a jigsaw on to a table and put it together manually without the

---

[h] Ham Smith suggested that *Hi* would be an excellent test case, given its relatively small size and that its nucleotide composition was similar to that of the human genome. Ham was already thinking that completing a bacterial genome sequence was a logical stepping stone towards the 'holy grail' of sequencing the human genome.

[i] There are two principal methods for this fragmentation and sequencing process. "Chromosome walking" requires first mapping the genome by constructing an ordered set of DNA fragments, as Fred Blattner had done for *E. coli*. These fragments are then sequenced one by one. In contrast, shotgun sequencing uses random fragments without any prior mapping. It is faster but a more complex process.

[j] A physical map of a piece of DNA depicts the location and distance between precise landmarks — for example those defined by cutting sites for restriction enzymes. In Chapter 9, I likened this to the map of a bus route with its designated 'stops'.

help of the picture on the box. The phage viral genome was relatively small, but with the power of TIGR's highly sophisticated computer programs, Ham thought he could use the shotgun approach[k] for the much larger (two million bases) *Hi* genome. An application to fund the project through the NIH was rejected, but Ham and Craig Venter went ahead anyway using TIGR money.

In the summer of 1994, Ham telephoned me. In stark contrast to his sombre mood of the previous year, he was evidently excited. Typical of his honesty and generosity, he told me that he regretted being so negative when I'd initially raised the possibility of sequencing the *Hi* genome the year before. He urged me to call Craig and impress on him how the TIGR sequence data could be used to further our meningitis research in Oxford. The outcome of my telephone discussion with Craig exceeded all my expectations. Evidently, my enthusiasm for the biomedical application of the bacterial sequencing project had touched a sensitive and sympathetic response in Craig, a legacy perhaps of his nightmarish experiences as a paramedic in Vietnam. Craig agreed to send us their DNA sequence data, compacted and encrypted on *floppy discs*. Using standard computer software, we could use the TIGR sequences as *DNA probes*[l] to search for matches to endotoxin genes logged into publicly available databases.[m] Using a computer rather than experiments in the laboratory was a completely new way to identify novel *Hi* genes. But these *in silico* findings[n] had to be confirmed and extended through laboratory experiments, work that needed months to obtain the required results. However, the prospects for making rapid progress were now within our grasp — although I anticipated there would be a long wait until the TIGR team delivered their promised sequence data.

---

[k] He had worked out, using simple programs on his home computer, the number of sequences and physical gaps that were expected using this approach. His "back of the envelope" estimates were remarkably accurate.

[l] Technically known as a hybridisation probe. This is a small fragment of DNA or RNA which can be labelled (using radioactivity or luminescence). Because of the strict pairing rules of nucleotides (adenine to thymine and guanine to cytosine), the labelled probes can be used to detect DNA or RNA through the binding of sequences of complementary nucleotides.

[m] GenBank or European Molecular Biology Laboratory.

[n] Meaning performed on a computer or via computer simulation. This use of pseudo-Latin (coined in 1987) was a tongue-in-cheek allusion to commonly used terms such as *in vitro* (outside living organisms), *in vivo* (within living organisms).

Meantime, using *Hi* gene sequences as DNA probes, we identified five matching genes for meningococcal endotoxin biosynthesis.[o] A few weeks later at an international meeting in Winchester[p] (1994), I found out that Emil Gotschlich from the Rockefeller[q] had identical data on the endotoxin genes of the gonococcus (*Neisseria gonorrhoea*), a bacterial species closely related to the meningococcus.[r] The events leading to his discovery were a good example of serendipity in science — and how chance favours the experienced and alert mind. Emil had been trying to clone the gene for a membrane protein, the main target for protective antibodies of the outer membrane vesicle (OMV) vaccine described in the previous chapter.[s] To his disappointment, the DNA sequence that he had pinned his hopes on being the sought-after gene turned out to have nothing to do with the membrane protein. So, what could it be? Puzzled, Emil ran a computer search and found that his unknown DNA was a close match to the *Hi*-b endotoxin gene sequences that my former post-doctoral researcher (Jeff Weiser) had discovered several years previously in my Oxford laboratory. Unwittingly, we had been the architects of our own downfall;[t] our endotoxin sequence had been put in the public data base and Emil had found it was a match to his unknown DNA. Unfortunately for us, he had already submitted a manuscript that was published months before ours.

Being beaten in the race to publish a novel finding is always a bitter pill to swallow. Scientists are competitive and like to be the first to publish a discovery. In most cases, being "scooped" does not have lasting consequences. What people usually remember is that, within a short time of each other, two research

---

[o] As in *Hi*, there was a reversible genetic switching mechanism, controlled by repetitive DNA, that resulted in variable expression of endotoxin. Both *Hi* and the meningococcus had evolved similar mechanisms to evade recognition by immune responses during the infectious process.

[p] The 1994 biannual International Pathogenic Neisseria Conference. Michael Jennings in my laboratory had done the research.

[q] The pioneer of the polysaccharide meningococcal vaccines. See Chapter 5.

[r] The gonococcus causes the sexually transmitted disease, gonorrhoea, often known as the 'clap', possibly derived from the French word for a brothel, *clapier*.

[s] This outer membrane protein was known as PorA.

[t] To add to the intrigue, knowing that Jeff had worked on endotoxin in Oxford, Emil told him that on no account would he be allowed to continue this research at the Rockefeller. How ironic that, based on the sequence data that Jeff had deposited in the public database, Emil identified the enzymes (glycosyl transferases) for endotoxin biosynthesis. So much for Emil's antipathy to research on endotoxin!

groups independently identified the same gene. Being scooped by Emil was annoying at the time, but in the grand scheme, a minor setback. But when a discovery is high-profile, being first can be seismic in its importance. The rivalry between Pasteur and Koch, described in Chapter 2, is a good example. Charles Darwin suffered agonies when he received a letter from the naturalist Alfred Wallace outlining a theory of evolution that captured the essence (but not the detailed arguments) that had taken the former almost two decades to consolidate. The importance of winning the race to solve the structure of DNA, captured so vividly by Jim Watson in the *Double Helix*, was enormous. In another context, two French scientists, Luc Montagnier and Francoise Barre-Sinoussi, received the Nobel Prize for their research on identifying the virus causing HIV, although the American Robert Gallo considered that he should have been credited as a co-discoverer. His very public claim to fame was not shared by the Nobel Committee. In another prestigious scientific race, Venki Ramakrishnan describes in his memoir (*Gene Machine*) the rivalry surrounding the research to elucidate the structure of the ribosome, the organelle within cells responsible for making all our proteins, for which he received the 2009 Nobel Prize in Chemistry.

What was more serious was that Emil had filed a patent — and *this* did have some potentially serious implications. It meant that he owned the intellectual property — a potential problem if we wanted to develop a vaccine based on the genetics of endotoxin. My discussions with the Oxford University technology transfer experts did nothing to alleviate my concerns. But to add insult to injury, their further enquiries found that Jan Poolman had also filed patents protecting other aspects of meningococcal vaccine development. In truth, I had been asleep at the wheel and completely naïve. There was little to be done and so I went ahead with the science anyway. I was advised that if a promising vaccine candidate emerged from our research, then likely there would be ways to protect it. The biochemistry of endotoxin was complicated, so would probably allow us to forge a legal passage through the intellectual property jungle.[u]

---

[u] Patent lawyers are adept at defining novelty — in legal terms designated as a discovery that is *non-obvious* to someone skilled in that area of research. The description must be able to counter the argument that the research is not novel (designated *prior art*) and therefore not eligible to be patented.

Meantime, Emil's and our research provided insights into the structure of endotoxin whose sugars, projecting outwards from the bacterial surface, were potential targets for protective antibodies. But genetic switching, driven by the effects of DNA repeats, had the potential to vary the presence or absence of the sugars of the endotoxin molecule and compromise the binding of antibodies. The reductionist power of genetics, however, allowed us to identify a region of the endotoxin molecule that was conserved and therefore a promising basis for a vaccine (see Figure 17.1).

I discussed this idea for a vaccine with a colleague, Rino Rappuoli, whom I had first met some two years earlier at a SAGE[v] meeting at the WHO Headquarters in Geneva. As the lead research scientist of Chiron Vaccines, based in Siena, he had made a presentation on their plans to develop meningococcal conjugate vaccines.[w] After the formal meeting, we all went to the "old city" to have dinner during which Rino and I chatted about future challenges, especially a vaccine against Meningococcus B. The upshot of our discussion was that he invited me to visit him in Siena and give a lecture to the Chiron Vaccine Research Group. It was the beginning of what would become a close, long-term collaboration, as I describe in the following chapters.

References

[1] Heather, J.M. and Chain, B. The sequence of sequencers: the history of sequencing DNA. *Genomics*, 2016, 107:1–8.

[2] A Life Decided. *My Genome My Life*. J. Craig Venter. Penguin. Allen Lane, 2007.

[3] Mukherjee, S, Gene, an Intimate History. *Vintage*, 2017, 309.

[v] Strategic Advisory Group of Experts.

[w] See Chapter 16.

Chapter 18

Siena and Vaccine Research

I have vivid memories of my first visit to Siena where I fell in love with the city and its stunning surrounding countryside. I recall wandering down *Via Franciosa* towards *Porta Camollia*, a triple-arched northern gateway[a] to the centre of the city, with its *bas-relief* bearing a Medici heraldic shield with the inscription *Cor magis tibi sena pandit*.[b] The city also provides a physical reminder of the brutal impact of infectious diseases. Its magnificent cathedral remains unfinished, a legacy of the fourteenth-century plague outbreak, the Black Death, from which 60% of Siena's population perished. Since medieval days, when pilgrims to Rome passed through Siena, visitors pour into the walled city whose cultural legacies continue to be a major tourist attraction. These medieval traditions include the seventeen *contrade*,[c] each of which lays claim to its own sector of the city. On festival days, members parade — dressed in their distinctive colourful doublets and hose — accompanied by loud drumming. Twice a year, the *Piazza del Campo* becomes a racecourse as horses and jockeys, representing each of the *contrade*, compete for the honour of winning *Il Palio*.

---

[a] The name of the gate comes from a soldier, Camulio, sent by Romulus, founder of the Eternal City, to capture his grandsons, Senio and Ascanio. But Camulio became distracted from his errand, and Porta Camollia, leading to Florence, was central to the defence of the fledgling town he built instead. Siena prospered, becoming a wealthy city of merchants and bankers and rivalling its neighbour Florence in prestige.

[b] "Siena opens its big heart to you."

[c] A *contrade* is a district, or ward, within an Italian City of which the 17 associated with Siena are the most well-known because of the famous *Il Palio*, run twice a year, in which representatives on horse-back compete to win this emotionally charged and dangerous race.

On a glorious early-spring day, Rino Rappuoli invited me for a typical Sunday Tuscan meal with his family who for many generations grew crops, olive trees and vines as well as raised sheep and cows. When not at school as a youngster, he worked from early morning to late at night in the fields. His father, a wine maker in the Chianti region, introduced me lovingly to Brunello di Montalcino, one of the great wines of the world. I could not help but wonder how it had come about that Rino, lacking any scientist role models within his close or extended family, would later become an inspirational and global leader in vaccines.

The day after my memorable Tuscan family meal, I met with Rino's research team to hear about their progress towards developing A and C meningococcal conjugate vaccines.[d] However, the main discussions centred on the complete failure of MenB conjugates to induce antibodies. Rino was adamant that a novel approach with new technology was needed, echoing Sydney Brenner's adage that:

> Progress in science depends on new techniques, new discoveries and new ideas, probably in that order.

During my lecture later that day, I summarised our research on using a region of the endotoxin molecule as a possible vaccine antigen. I hoped to attract some funding from Chiron, but Rino made it clear there was no prospect of research support unless I could provide compelling evidence that antibodies were protective. It was a challenge I needed to discuss with my collaborators at the National Research Council (NRC) in Canada.

Ottawa in winter, the river frozen solid and armadas of snow ploughs clearing the roads. The wind chill factor is way below zero and I have arrived with all the wrong clothes and footwear. Waking early in the morning after my transatlantic flight, I put my finishing touches to a talk that I have prepared for the NRC research team. The Oxford and NRC research groups have complementary skills. Our expertise in genetics aims to identify the genes required to make the endotoxin molecule. By making mutations, we have isolated the sugar components that we hope will be the basis of a vaccine.

---

[d] Chapter 16 describes the Chiron MenC vaccine.

Using "high-tech," state-of-the-art analytical instruments, NRC have the expertise to detail the spatial arrangement of the endotoxin molecule that projects outwards from the surface of the meningococcus (see Figure 17.1). It's an exciting collaboration and soon after my return to Oxford there is clear evidence that antibodies can target the surface exposed regions of the endotoxin molecule. Importantly, NRC have made an antibody that kills meningococci in a laboratory assay — Rino's *sine qua non* for obtaining funding for the project. I wanted to go one step further and show protection in animals, but it would be several months before these important studies could be done. I was not too upset at the delay because, in the meantime, the first tranche of *Hi* genome sequence data had arrived in Oxford from TIGR and we were going to have our hands full to take advantage of this amazing opportunity.

Based on the innovative application of the shotgun method, the two million paired nucleotides of the *Hi* genomic DNA had been sheared into thousands of small pieces. Ham had insisted on doing this himself to ensure that every region of the genome was represented. He understood the huge importance of random fragmentation. To ensure complete coverage, it was necessary to sequence around 60 million nucleotides (30 times the size of the genome), but this statistic was only valid if the genomic DNA fragmentation was random. DNA was extracted from billions of bacterial cells and purified to yield a clear viscous fluid that filled a small plastic tube smaller than the size of my little finger. To fragment the DNA, Ham tried several methods and found that the ideal method was to atomise it using a perfume spray bottle that had been discarded by his wife. This reliably produced random fragments of the correct size. After several iterations of this process, the thousands of pieces of DNA were cloned into *Escherichia coli* bacteria that could be grown in culture to amplify the DNA and obtain enough material for sequencing. TIGR then used their sophisticated computer programs to search the thousands of DNA fragments to identify overlapping sequences so that the entire genome could be pieced together.

While all this was going on, my Oxford team were also using computer programs to look for matches between TIGR *Hi* DNA sequences and the hundreds of genes for making endotoxin or other polysaccharides in other species of bacteria that were available in the publicly available databases.[e] The search for matches to our *Hi* DNA sequences were spectacularly successful; within a few days, we had identified many possible genes for *Hi* endotoxin. Using classical genetics, we made mutations in these different genes to prove that they were involved in making *Hi* endotoxin and test what effects these alterations had on causing meningitis. Whereas the computer searches had taken only days, the laboratory experiments took much longer. The eventual outcome was hugely rewarding. More than 20 new *Hi* endotoxin genes were identified — more progress in a few weeks than we had made over several years. It was a good example of how molecular biology and genomics had changed research on bacterial pathogens.

I called Craig to tell him our exciting news. By chance, Ham was visiting TIGR and I ended up talking with them both. TIGR was close to completing assembly of the *Hi* genome sequence, the remaining challenge being to work out how all the DNA fragments fitted together to form a circular genome. Using the computer programs, the TIGR team had reduced 25,000 fragments of DNA to about a dozen larger fragments. Only a few gaps remained; the dream of the completely assembled *Hi* genome sequence, a milestone in biology, was now just a matter of time. I suggested that we plan a major international meeting to discuss the revolutionary impact of genome sequencing of bacteria. The TIGR team would present the complete genome sequence of *Hi* and my laboratory would show how this information could impact public health through a better understanding of a major bacterial pathogen responsible for meningitis. Ham and Craig were enthusiastic, so I called one of the senior scientists at the Wellcome Trust (WT) who encouraged me to submit a formal application to obtain funding for the meeting.

The process proved to be less straightforward than I had hoped. The *Hi* genome sequencing project had not yet been finished and the meeting was scheduled to happen before the publication of any peer-reviewed scientific article. The WT rightly questioned whether the project would be finished on time and some reviewers even cautioned that perhaps the whole project was "hyped up." One senior member of the WT scientific staff wanted to see

---

[e] When scientists working on other bacteria publish results on genes sequences, they are submitted to organisations such as the European Molecular Biology Organisation so that other researchers can quickly access this data to help their research.

a compact disc of the sequence prior to giving the go-ahead. Others voiced concern that the meeting might be compromised by commercial secrecy agreements[f] or that Craig would not attend the whole four days, one of the sacrosanct conditions laid down by the WT concerning attendance of their *Frontiers in Science* meetings. But many influential scientists gave enthusiastic support and, to my relief, funding for the meeting was finally approved.

---

[f] Although TIGR was a "not-for-profit" company, it had received substantial funding for its research from Human Genome Sciences, a commercial company that aimed to develop protein and antibody drugs.

Genomics and the Wellcome Trust

In April 1995, 50 of the world's leading scientists assembled in the heart of the Cotswolds, for a meeting that would usher in a new epoch in biology. Craig Venter described how the TIGR team had sequenced the thousands of pieces of DNA and used computer programs to assemble the complete bacterial genome of *H. influenzae* (*Hi*). As he talked, a large video screen scrolled through the entire sequence, timed to coincide with the length of his talk, a typical piece of Venter showmanship. Here was the genetic information for making 1800 or so proteins that underpinned all the activities required for independent survival and reproduction of a cell. The atmosphere in the meeting room was electric. Many participants hadn't known or were still sceptical that the DNA sequence of a bacterial genome had been completed. The TIGR scientists had set up several computers in an adjoining room where, after the formal talks were over, participants could inspect the complete genome sequence for themselves. Many were still there in the early hours of the next morning as they identified, often for the first time, the DNA sequences of genes that were crucial to their research. This was also a new way to identify virulence factors and novel targets for treatment or prevention of *Hi* infection.

After presentations from the TIGR scientists, it was Oxford's turn to present our findings on how genomics had identified the 25 novel genes for the endotoxin molecule. Their location on the circular genome had been mapped and the role of the component structures of the endotoxin molecules encoded by these genes had been characterised in an animal model of meningitis. The several regions of repetitive DNA were another striking feature, providing insights into the genetic basis of bacterial variation in these surface structures, a key mechanism allowing bacteria to adapt to humans and evade immune responses.

Venter quickly recognised the implications of using complete genome sequences to advance understanding of pathogens. As he lectured across the globe, he included slides showing how DNA repeats provide a mechanism for bacteria to alter their surface structures, adapt to their hosts and evade immune clearance mechanisms. For many scientists, the *Frontiers in Science* meeting provided the first intimation of a revolution in biology that was going to change the *modus operandi* of science. In the field of infections, it was a cost-effective way to acquire new knowledge on epidemiology, pathogenesis, antibiotic resistance, diagnostics and prevention. But, in its broader impact on biology, the sequencing of different life forms would result in a radical re-interpretation and understanding of the tree of life. Charles Darwin's seminal book *On the Origin of Species* (1859) was published around the time that the "germ theory" of infectious diseases was formulated. But his theory of natural selection was developed without any major consideration of microbes.[1] Rather, it was based largely on his observations on life forms that evolved relatively recently, including various flowering plants, worms, birds and domesticated animals. For the next 100 years, scientists wrestled with the question of how to integrate bacteria into the tree of life (Figure 19.1). In the late 1970s, Carl Woese

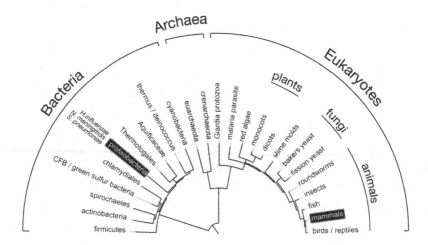

Figure 19.1   The Tree of Life. The first life forms were bacteria that appeared on planet Earth some 3.8 million years ago. If the time span of the history of life is considered to be a 12-hour clock, *Bacteria* occupy the entire 12 hours, the first *mammals* evolved less than 10 minutes before 12, humans (see mammals on right) a few seconds before 12. The bacterial species that cause meningitis are shown on the left (belonging to the proteobacteria). Their co-evolution with humans suggests a relatively recent emergence.

proposed the radical idea that bacteria occupied nearly the entire tree; there was just one branch for all animals and plants, a hugely controversial theory. The availability of complete genome sequences of all domains of life forms, of which the *Hi* bacterium was the first example, cemented the correctness of Woese's "Tree of Life." The "shotgun" sequencing and the use of sophisticated software computer programs to assemble the complete bacterial genome was also the basis of the strategy that would be adopted by Craig Venter's Celera team to sequence the human genome.

At the conference dinner that evening, Craig presented me with what in those days was a familiar small square object, the now obsolete floppy disc,[a] containing the complete genome sequence of *Hi* signed by each member of the TIGR project team. "I'm expecting a vaccine very quickly now that we have provided you with all this information," Craig challenged.

"Where've you been?" I responded. I was chuckling because Craig was clearly unaware that the *Hi*-b conjugate vaccines were already part of the routine infant immunisation programmes in many parts of the world. "We don't have a vaccine against MenB and I want you to do that genome next," I told him. Craig and I agreed to continue our collaboration and extend the *Hi* project to include sequencing a strain of meningococcus isolated from a child with meningitis in the Stroud outbreak of MenB meningitis (see Chapter 14) that had occurred in the 1980s only 20 miles from the conference centre.

In the summer of 1995, a few months after the Genome conference meeting in the Cotswolds, Craig Venter and I arranged to meet with the Wellcome Trust's lead scientist on genomics, to discuss sequencing the genome of MenB. The informal get-together was a dreadful anti-climax. Michael Morgan hadn't attended the *Frontiers in Science* meeting and seemed unaware of its impact. It was a good example of how scientists can live in parallel universes. His attitude was also indicative of a mind-set at that time. Although the Wellcome Trust (WT) was hugely committed to the human genome project, sequencing of bacterial genomes was considered small beer and wasn't on his radar screen.

Despite this disappointing lack of interest from the WT, Craig made a commitment to me that TIGR would begin the MenB project. But he made

---

[a] Used as a magnetic storage device. The cassette provided access to almost 3 MB of data when inserted into a disc drive on a computer. In 1995, it had only just become available as a standard piece of hardware.

no secret of the fact that he was short of money and any long-term commitment depended on my getting secure funding. In subsequent discussions, other scientists at the WT were more enthusiastic about my plan to sequence a MenB genome, and within months we had encouraging preliminary data. Craig had kept his promise; TIGR had sequenced 400 fragments of DNA, each about 500 nucleotides in size, representing around two-fold the genome coverage. From this, we identified a novel gene[b,2] for a protein on the bacterial cell surface, a proof of concept for identifying novel vaccine candidates. In March 1996, I submitted an application to sequence the MenB genome. A few weeks later, I was thrilled to receive a provisional award letter. The WT had given the project a very high priority for funding, but confirmation depended on clarification of the "not-for-profit status"[c] of TIGR. This sounded to me as if there was just some rubber stamping before we got the go-ahead.

Craig and the TIGR team were excited, but celebrations were premature. What followed were months of legal wrangling, confusing communications and ultimately disappointment. Nobody at the WT doubted the merit or the importance of the science, but several legal and political issues came to the fore. The legal concerns were that TIGR had received funding from Human Genome Sciences (HGS), a biopharmaceutical company launched in 1992 of which Craig was one of three founders. Although TIGR was a "not-for-profit" organisation, HGS had a clear intent to seek patents on DNA sequences that had therapeutic potential. My proposal that all the intellectual property would be assigned to the University of Oxford had met with no objection from Craig and TIGR's lawyers. But the reality was that the WT was belatedly having second thoughts. It had put millions of pounds into setting up a Pathogen Sequencing Facility at the Sanger Institute. Ironically, this had come about through my recommendation to the Governing Body as part of my summary of the *Frontiers in Science* meeting. But early in 1996, the WT team had neither the expertise nor the dedicated facilities to sequence, assemble and annotate bacterial genomes. TIGR was in a league of its own at that time,

---

[b] Later work by two of my post-doctoral scientists, Michael Jennings and Ian Peak, resulted in this antigen being licensed by GlaxoSmithKline.

[c] A non-profit organisation is a business that has been granted tax-exempt status because it furthers a social cause and provides a public benefit. All profits must be used to pursue the organisation's objectives.

unique in its degree of expertise and with a proven track record to tackle the MenB genome. As it turned out, the Sanger scientists subsequently made astonishing progress on sequencing pathogens, and by 1998 they had completed the genome of the TB bacterium (*Mycobacterium tuberculosis*). But the pace of progress in genomics was such that delaying the project for a year or so — even a few months — was problematical. I wanted to start the MenB project straight away.

There was a further complication that I had not anticipated. The WT had decided on a policy that their DNA sequence data should be made publicly available on an immediate basis. This aimed to facilitate scientific progress and deter the patenting of genes, an issue that had caused much controversy ever since the explosive impact of Craig Venter's research at the National Institutes of Health.[d] In 1996, a meeting was held in Bermuda that championed the principle of immediate release of sequence data. This joint policy statement by the WT and the National Institutes of Health was controversial. Craig and other scientists argued that what became known as the "Bermuda Proposal" was a precedent for a completely new way of doing science. With the *Hi* genome, TIGR had completed the sequencing, analysed the data, published it in a peer-reviewed top-flight scientific journal and only then released the data into the public domain. Craig was hostile to the immediate release policy. "… Can you imagine Fred Sanger posting on a wall outside his office their results on the sequencing of insulin for other groups to use? I mean it sounds absurd, right?" Craig exploded when asked about the reasons for his opposition to what he characterised as "nightly dumping of crude, unedited, often inaccurate DNA sequence data on to publicly available servers." These differences in opinion had morphed into an ugly feud.

In December 1996, nine months after submitting my application, I received a letter from the Director of the WT, Bridget Ogilvie, informing me that:

> *The Scientific Committee has recently considered the circumstances surrounding the conditional award to the University of Oxford, on which you would be the Principal grant holder, to facilitate the sequencing of the genome of Neisseria meningitidis at The Institute of Genomic Research (TIGR) under*

---

[d] See Chapter 17.

the direction of Dr Craig Venter. The Committee has decided that it will not be possible, despite considerable efforts, to resolve the present impasse, which it considers to be a direct result of the contractual relationship between TIGR and Human Genome Sciences (HGS) Inc.

This was a bomb shell. I wrote immediately to request a meeting to discuss more fully the reasons behind the decision, but my letter went unanswered and nobody at the Trust would agree to discuss the decision with me. The relevant Committee was not identified and more likely referred to informal discussions among key WT personnel. I was up against a brick wall. More than 20 years later, in doing research for this book, I requested release of all the related correspondence — to which the WT readily agreed. Much was not shared with me at the time, especially correspondence with TIGR lawyers. I can understand in retrospect how the issues of TIGR's not-for-profit status, their links to Human Genome Sciences and the clash over policies for data release created many complications. It was a salutary example of the complex impact on science of politics and inter-personal interactions.

Fortunately, it was not the end of the story. Within months, the MenB project would be rescued through a collaboration with Rino Rappuoli and Chiron Vaccines in Siena. It would pave the way for a completely new approach to vaccine development.

References

[1] Moxon, E.R. Darwin, microbes and evolution by natural selection. *Advances in Experimental Medicine and Biology*, 2011, 697:77–86.

[2] Peak, I.R. et al. Evaluation of Truncated NhhA Protein as a candidate meningococcal vaccine antigen. *PLoS One*, 2013, 8(9):e72003. doi: 10.1371/journal.pone.0072003.

The Last Frontier: A Vaccine Against Meningococcus B

In the autumn of 1996, I was asked by Richard Horton, editor of *The Lancet*, to be the scientific adviser for a series of articles[1] on recent advances in the field of vaccinology. My own contribution was an article in which I discussed how "… complete genome sequences provide a catalogue of the genes for every virulence factor and potential immunogen from which to select vaccine antigens." This was the first description of what is now a routine approach to developing vaccines — the most recent example being the *spike protein* of COVID-19.[a,2]

Of course, the publication of the *Hi* genome sequence had galvanised huge interest among microbiologists everywhere and its potential for accelerating vaccine development was of major interest to the big pharmaceutical companies. One of the first commercial technology platforms was set up by a US company[b] who sequenced the genome of the bacterial pathogen *Helicobacter pylori*, the cause of duodenal ulcer and gastric cancer. Their data — although never made public — were licensed to the Swedish pharmaceutical company Astra[c] AB as part of a $22 million agreement. Genomics was big business.

In Siena, Chiron scientists were planning to exploit their DNA sequencing facilities to take a genomic approach to developing a MenB vaccine. At the time, the company was undergoing a massive shake-up. Their Chief Information Officer,[3] a former microbiologist, had a special interest in genomics as a tool for drug discovery and had served on an advisory committee for the

---

[a] The use of genomics to discover vaccine antigens is now often called "reverse vaccinology" (see Ref. 2).

[b] Genome Therapeutics Corporation in Waltham, Massachusetts.

[c] Would become AstraZeneca.

(a)   (b)

Figure 20.1   (a) Craig Venter (1946–); (b) Rino Rappuoli (1952–).

US government. The Department of Energy Office had provided substantial funding to TIGR for some of its early bacterial genome sequencing projects. From first-hand knowledge of TIGR, it was evident that Chiron's modest "in-house" sequencing capacity would be wholly inadequate to take on the MenB genome project and a meeting was set up between Rino Rappuoli and Craig Venter at the TIGR headquarters in Maryland.

TIGR had deposited the limited DNA sequence data from the stalled Oxford project on the GenBank database. It was a start, but much more genome sequencing was needed. When Rino arrived at TIGR for the pre-arranged meeting, he found that Craig's priorities had changed. The news was just out that Jim Watson had been awarded the National Medal of Science by President Clinton for his contributions to launching the Human Genome Project. Craig was frantically grappling with ideas as to how TIGR could speed up the efficiency, accuracy and scope of its technology to sequence the human genome of three billion nucleotides. Craig was convinced that he could apply the shotgun approach that had been so successful for *H. influenzae* (*Hi*) to shorten the time and costs (by hundreds of millions of dollars) of sequencing the human genome. To this end, he was planning much more ambitious sequencing projects than those of bacteria, for example the genome of the fruit fly with its 170 million nucleotides compared to the 2 million nucleotides of bacteria such as MenB.[d] The fruit fly, *Drosophila melanogaster*, had been chosen early in the twentieth century by T. Dobzhansky and E.B. Morgan as a tractable life form on which to investigate genetics. Their large chromosomes are easily visualised in their salivary glands making them ideal for laboratory research. Mutations, for example those that change eye colour, are easily identified. Their quick reproduction rate allows scientists to observe the effects of these mutations. Of the 289 human genes known to be involved with disease, more than 60% have a related counterpart in the fly.

Despite Craig's reluctance to undertake sequencing of another bacterial genome, Rino's advocacy of the importance of meningitis as a major public health problem won Craig over, and in December 1997 a deal was signed in which TIGR agreed to sequence the genome of MenB isolate in exchange for royalties if a successful vaccine ensued.

A few weeks later, after the ink was dry on the Chiron–TIGR agreement, I received a telephone call from Rino asking me if I was willing to collaborate. It came as a complete surprise. I had no idea that negotiations had been going on between Chiron and TIGR, but here was an opportunity to revive the MenB genome project with TIGR in collaboration with the very talented group of Siena-based scientists, a partnership that would eventually span more than two decades.

Over the next several weeks, Rino and I had discussions in Oxford and Siena. Both of us were passionately committed to developing a MenB vaccine, the "last frontier" in preventing the major causes of bacterial meningitis. In addition to using genomics to identify novel protein antigens, I wanted my group in Oxford to continue research on the MenB endotoxin vaccine. The two projects were complementary as the genome sequence would be a great help in identifying the genes for making endotoxin — as we had already shown for *Hi*.

Rino and I discussed the issue of intellectual property rights and funding. It was an awkward discussion because I was now much more aware of what was at stake. It was immediately evident to me that Rino was uncomfortable as he admitted that Chiron had filed patents on the use of the MenB genome

---

[d] Craig Venter made decisive contributions to the Human Genome Project. Three years later, in June 2000, when President Bill Clinton strode into the East Room of the White House to announce the completion of the Human Genome Project, he was followed closely by the US heroes of the hour: Francis Collins, Jim Watson and Craig Venter.

sequence to identify vaccine antigens.[e] For a few minutes I thought that we were on the verge of an unpleasant disagreement. After all, Oxford had initiated the project and had provided The Institute for Genome Research with the DNA and the bacterial isolate that was being sequenced. Yet, the business transactions and filing of patents had been done without my knowledge and Craig had never contacted me to discuss what was going on. I was still angry and disappointed over the debacle with the Wellcome Trust.

But Rino had shrewdly anticipated my concerns and countered them by offering to provide substantial research funding for my laboratory on an immediate basis. From my point of view, this had its advantages. The MenB genome project to develop a vaccine might eventually result in a royalty stream,[f] a proportion of which would come to my department through the University of Oxford if I was a named collaborator. But these revenues would not materialise, if at all, for more than a decade, probably longer. I needed immediate funding to support the Oxford scientists and the laboratory investigations that were necessary for the MenB vaccine project, including the costly experiments to find out if antibodies to endotoxin were protective in animals.

In the next decade, my laboratory would receive more than a million dollars of research funding from Chiron, the first tranche of which became available within months of my meeting with Rino. It seemed a good outcome. Importantly, I had realised that developing a MenB vaccine of such complexity was absolutely dependent on having a commercial partner. In turn, Rino and the Chiron team needed the experience of my Oxford laboratory in what was for them the relatively new field of genomics. For example, we had compiled a collection of hundreds of carriage and disease isolates of MenB and other meningococci. To make an effective vaccine, the components had to induce antibodies that would protect against the highly variable protein or endotoxin structures expressed on the surface of MenB isolates that caused meningitis. Our carefully assembled collection of meningococci was a key resource. Overcoming

---

[e] The patenting of genes is highly controversial but, for a commercial company, the investment of millions of dollars in developing a vaccine could only be justified if it owned the intellectual property.

[f] Eventually, the revenues from the MenB vaccine would amount to billions of dollars. I think it can be justly noted that whatever merits I may have as a scientist, it can be concluded that I am a lousy businessman!

the challenge of variation, the omni-present metaphorical "elephant in the room," is one of the most crucial challenges in making an effective vaccine.[g]

By 1998, the TIGR–Chiron–Oxford MenB genome vaccine project was moving forwards apace. Chiron's research team in Siena had already searched the limited DNA sequence data from the earlier Oxford–TIGR collaboration,[h] while the TIGR team set about the extensive sequencing that would hugely increase the amount of data from which to search for vaccine antigens. It wasn't necessary for the genome sequence to be completed prior to searching for novel vaccine antigens. But assembly of the entire two million nucleotides that make up the circular genome, the natural state of the DNA within the bacterial cell, was needed to make sure that no potentially relevant vaccine antigen had been overlooked. It was the completeness of this search for vaccine antigens that made the genomic approach so appealing.

The DNA of the MenB genome contains around 1,300 genes of which only a minority code for proteins expressed on the bacterial cell surface, accessible to antibodies and therefore credible as vaccine antigens. The first challenge was to identify these candidate proteins using what computer scientists call bioinformatics — the use of mathematics, statistics and probability theory to devise computer software programs (algorithms) that predict biological functions. DNA contains the code that specifies the sequences of the building blocks (amino acids) of all the proteins of the bacterium. By integrating knowledge obtained over decades from biological observations and experiments on the huge number of known proteins from all domains of life, predictions of the structure and functions of all these new proteins in the MenB genome could be obtained. Although the state of the art of bioinformatics in 1998 was not a patch on what it is today,[i] it was good enough to identify which proteins were likely to be located on the bacterial cell surface. These were then subjected to further

---

[g] See Chapter 21 for further discussions on microbial variation and its implications.

[h] There were also data from a sequencing project on the gonococcus being carried out by a team from the University of Oklahoma.

[i] Today, sophisticated software and dedicated suites of programs exist to accurately predict a protein's cellular localisation and potential biological function. This was not the case 20 years ago. In the late 1990s, interrogation of sequence data was in its infancy and the utility of many of the algorithms was not validated. But it was a new way of doing science and the MenB project would prove to be the first example of developing a vaccine whose main components were identified through genomics.

rigorous testing in the laboratory. Although some predictions proved spurious, the analysis identified hundreds of credible antigens for further investigations.

For the Chiron scientists, using genome sequence data as the starting point for identifying vaccine antigens was a totally new concept, fundamentally different from what they had been accustomed to. Many were sceptical; instead of analysing and purifying bacterial components at the laboratory bench, the scientists were sitting in front of a computer using bioinformatics to identify hundreds of novel proteins from the conveyor belt of the genome sequence; it was not at all the kind of science that the team had been trained to do. But Rino's vision and determination prevailed and, as often happens in science, the interactions of the different personalities and expertise was catalytic. By the time of the first joint meeting of Oxford, TIGR and Chiron scientists in Siena, there was a mood of excitement and optimism. Nor was the collaboration hindered by the delights of Tuscan food and wine enjoyed in one of Siena's traditional restaurants located close to the Piazza del Campo after the many hours of research discussions.

From 570 candidate proteins, the coding DNA was cloned into *E. coli,* the workhorse laboratory organism used for virtually all recombinant DNA research. Each *E. coli* clone, expressing the DNA for one candidate antigen, was used to immunise mice. After testing the mouse sera, the tally was reduced to less than a hundred and, after more stringent testing, there were twenty-eight promising, novel proteins. The genomics approach had delivered in a spectacular manner (Figure 20.2). Over several decades, classical laboratory methods had identified only a handful of promising

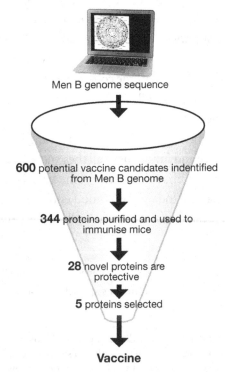

Men B genome sequence

600 potential vaccine candidates indentified from Men B genome

344 proteins purified and used to immunise mice

28 novel proteins are protective

5 proteins selected

Vaccine

Figure 20.2 A genomic approach to a meningococcal vaccine.

MenB antigens, most of which were not taken forward for testing in the clinic because they did not induce antibodies capable of killing a sufficient proportion of the bacterial isolates from cases of meningitis.

In the meantime, the TIGR and Oxford scientists were trying to assemble the more than 2,000 base pairs of double-stranded DNA into the completed circular MenB genome. In the middle of 1999, word reached us through the "grape vine" that scientists from the Wellcome Trust's Pathogen Sequencing Centre were close to completing the genome sequence of a different meningococcus isolate.[j] My original application to the Wellcome Trust had been submitted in 1996 before the Pathogen Sequencing Centre had got into its stride. But by 1998, the Sanger Centre scientists had made astonishing progress. They had sequenced the genome of TB and a meningococcal genome was their next project. As James Watson wrote, "only saint-like minds can watch someone in the next lab race them from an experimental result and not get violently upset."[4] There is kudos in being the first to publish a breakthrough. My original application to the Wellcome Trust, initially successful but later rescinded, had delayed the project by 18 months. Now, the Sanger team were making rapid progress and I desperately wanted us to win the race. In December 1999, we submitted two articles to the prestigious journal *Science*, one on the complete genome MenB sequence and the other on how it had been exploited to identify novel candidate vaccine antigens. Both articles appeared in the same issue in March 2000.[k]

That morning, before I had set off to work, members of the press were on the doorstep of our house in North Oxford. It did not take long for me to find out why interest was so intense. "Meningitis boy died after 60-mile trip to three hospitals" was the headline of that morning's *Daily Telegraph*. In the *Daily Mail* there was a beguiling photo of a smiling, blond 2-year-old under the heading "Meningitis: Breakthrough over vaccine that might have saved this little boy." The article detailed how, alarmed by vomiting and purple spots on her baby-boy's chest, his mother had taken him to the local hospital. But there were no paediatricians and he was sent to another hospital 30 miles away,

---

[j] Meningococcus A, the cause of the epidemic meningitis that is especially common in Africa.
[k] The Sanger Centre published their complete genome sequence of MenA in *Nature* two months later.

under police escort. After giving him a blood transfusion, the staff decided that they were unable to provide the level of intensive care needed. Two hours later, his condition still worsening, he was transferred to a medical centre 90 minutes away. After four days on a respirator, he died.

"Meningitis is every mother's nightmare and my son should have been taken to the right hospital straightaway," the mother told the press. More than a year later, the boy's tragic death was debated in the House of Commons, highlighting the issue of whether his death would have occurred had there been a paediatrician at the local hospital. The parliamentary report concluded that earlier treatment might indeed have prevented his death. The boy's mother, a nursery nurse by profession, commented:

> "Doctors should have realised earlier that if his condition got worse, he would need paediatric intensive care. I want to prevent that happening again. I am not saying that Tyler would have survived if he had been taken to Nottingham straightaway, but by doing so in future cases it could save a child's life."

Many newspapers commented that immunisation would put an end to the possibility of these terrible tragedies; journalists wanted to know how long it would take to have a MenB vaccine. *The Guardian* splashed the headline: "Scientists Crack Meningitis Code" and incautiously ventured that "... trials of a vaccine based on the research could begin in 2001."

I knew it was many more years away and was relieved when a spokesman for the Meningitis Trust charity told the BBC that although this was a major step forward and very good news, there was still much work to be done. We had not even decided what exactly would be in the vaccine, let alone begun the many years of clinical trials that would be needed.

After the euphoria of the article published in *Science*, it was back to investigating which antigens should be used for the MenB vaccine. As each candidate was investigated in more detail to determine its protective efficacy, it became clear that no single antigen would provide adequate coverage against the wide range of MenB isolates. There were also practical constraints to consider; manufacturing costs, the technical challenges of making large amounts of each of the purified antigens and compliance with what would be acceptable to the two major regulatory authorities, the FDA (Food and

Drug Administration)[l] and the EMA (European Medicines Agency).[m] Taken together, these factors placed a limit on the number of antigens to three or four. There were also commercial imperatives. Chiron needed to recoup the research and development costs and therefore patent protection of the components included in the vaccine was important.

Given the public health importance of developing a vaccine against an appalling disease such as meningitis, being locked into these sorts of commercial imperatives will strike many as repugnant. But for many decades, only "Big Pharma" had been able to provide the combination of expertise and manufacturing capacity to comply with the regulatory requirements to develop and deliver safe and effective vaccines. Yet, pharmaceutical companies are hugely dependent on their shareholders and must remunerate them. Herein lies the complexity and tensions between commercial and public health imperatives.

There was a limit to the number of assays that could be done on the large number (28) of candidate antigens. Most antigens protected against only some, often less than the majority, of the panel of MenB strains. It was proving difficult to find a combination of antigens that would reach an acceptable level of protection. But what was an acceptable level?

To explore this, let's propose that a combination of four antigens targets 80% of MenB disease strains. Vaccines usually do not induce protective antibody responses in 100% of recipients. So now let's assume that 90% of those immunised make a protective immune response, then overall efficacy would be 72%[n] (80 × 90 ÷ 100). Acceptable but far from ideal and considerably less than many currently used vaccines. For example, the effectiveness of the MenC vaccine is greater than 90%. Could the addition of the (endotoxin) component improve the level of vaccine protection?

The animal experiments had been encouraging. Antibodies protected against a majority of MenB strains. The next step was to make the endotoxin

---

[l] The US government agency set up in 1906 to take responsibility for the safety of food, dietary supplements, human drugs, vaccines, blood products and other biologicals.

[m] The body responsible within the European Union for marketing authorisation of medicines for human and animal use, including vaccines.

[n] See Chapter 21 for the efficacy estimates of the MenB vaccine after its implementation in the UK.

sugars as immunogenic as possible by conjugating them to protein. The chemistry required was challenging even for our highly skilled collaborators at the National Research Council in Ottawa. But the method that worked best could not be scaled up to produce the amounts of vaccine required for clinical trials. Worse was to follow; alternative methods of making the endotoxin-conjugates did not induce protective antibodies. After more than ten years of research, we had failed to get the results we needed.

It is hard for me to convey the immensity of the disappointment we all felt. The Oxford–NRC team had put in so much work and had proof of concept that the surface exposed sugar component of endotoxin was a legitimate target for inducing protection, but we had failed to come up with a commercially viable method for making the required conjugates. It was bitterly disappointing. The reality of science is that too often the bottom line is not a triumphant coda but an anti-climax. It is realities like this that make me wish that I was writing a novel...

Meantime, in the spring of 2000, Rino had just returned from Australia where he had given a series of lectures on the MenB genome vaccine project. The research had attracted huge interest especially because of the outbreak of MenB meningitis in neighbouring New Zealand (NZ) that had been ongoing since the 1990s with devastating consequences. In a population of just four million, there had been more than 500 cases in 1999 alone, a twentyfold increase in life-threatening infections. The World Health Organisation (WHO) had flagged it up as a public health emergency and the NZ Ministry of Health desperately wanted a vaccine. Because the epidemic was caused predominantly by a single variant of MenB, it was amenable to using a vaccine made from outer membrane vesicles (OMVs), the strategy that had been successful in combatting the outbreaks in Norway and Cuba.

On the long flight back to Europe, Rino began to piece together a plan. If he could pull it off, Chiron could become the global leader in meningococcal vaccines. He reasoned that Chiron should make a bid for the competitive tender for an OMV vaccine that had been put out by WHO on behalf of the NZ Ministry of Health. This would not just respond to the major public health crisis. The MenB genome project had identified proteins that achieved only partial coverage against all MenB strains. The addition of OMVs to the formulation would increase its protective activity. But, given that Chiron had invested heavily in the new technology of genomics, persuading the company to develop an OMV vaccine was a daunting task.° Nevertheless, Rino convinced them that the OMV vaccine was not only of immediate importance for NZ but also, in the longer term, a vital component of a universal vaccine. Pragmatically, Chiron — the world's sixth largest vaccine manufacturer — had to outcompete GSK, the world's number two, as the chosen manufacturer of an OMV vaccine for NZ.

In February 2001, shock waves resonated through the pharmaceutical world. The panel of international experts assembled by WHO chose Chiron to provide the OMV vaccine for NZ. Like many others, I had assumed that the contract would go to GlaxoSmithKline (GSK) who, collaborating with the hugely experienced Cubans, seemed to be the obvious choice. But, Chiron Vaccines, partnered with the Norwegian Public Health Institute, had unexpectedly beaten out their higher-ranked rival. Their successful proposal was built on shrewd pricing, meticulous attention to the logistics of vaccine delivery, well-planned clinical trials and concrete proposals to overcome regulatory hurdles. There is also little doubt that the Chiron team had communicated a heightened sense of commitment that helped to carry the day.

Nothing exemplified this more than Chiron's clinical trials coordinator, Phillip Oster. Nobody told him to spend more time in New Zealand than in Italy. Most would have done the job by phone, but he was present at key meetings wherever they were held. A colleague teasingly asked him if he was working for Chiron or the New Zealand Ministry of Health. The scientists and public health workers could talk to him and found that he got to the heart of their problems. The nightmare of meningococcal disease had haunted New Zealand for more than a decade. Now they were in a partnership that could deliver a vaccine to put an end to the epidemic.

By 2002, the technical team in Siena were struggling to find ways to scale-up the procedures for the OMVs that had been developed in Norway. This was far from straightforward. In transferring the technology to Siena, not every aspect could be duplicated. The Norwegian MenB organisms used

---

° Rino was impressed by the successful MenC programme in the UK through which national immunisation had been implemented without any expensive, time-consuming efficacy trials. This precedent was an attractive model for NZ where rapid implementation of an OMV MenB vaccine was paramount.

to make the OMVs differed from those obtained from the NZ strain and required different conditions to grow the organism in a giant 50-liter fermenter. Indeed, the growth medium specified by the World Health Organisation was different from that used by the Norwegians. Quality assurance systems had to be established so the strength, sterility and consistency of each batch of OMVs could be repeated and the whole process had to be scaled up. Once again, the issue of making the vaccine in large quantities became a problem. Bacteria that had grown so well in the laboratory failed to do so in the large batch fermenters. This problem was eventually solved when it was discovered that the isolate from New Zealand was exquisitely sensitive to the concentration of iron in the growth medium. Once this was increased, normal growth occurred.

Now came small-scale clinical trials to ensure safety and appropriate immune responses. Finally, by July 2004 the OMV vaccine could finally be given to the target population: 1.2 million people up to 20 years of age. Its effectiveness could be assessed in a large-scale implementation programme.[p] The NZ public health team had done a remarkable job, but calculating vaccine effectiveness was complicated because, prior to the vaccine campaign, there had been a steady decrease in disease incidence between 2001 and 2004. Nonetheless, an undeniably accelerated decrease in the number of cases of sepsis and meningitis had occurred following implementation of the OMV vaccine. The extent of protection correlated with the number of vaccine doses. The vaccine had worked and there was much to celebrate. Three million doses of a vaccine, developed and delivered within three years (as compared to the usual 12–15 years), had been given to more than a million individuals, 80% of the target population. By June 2006, disease rates had fallen to pre-epidemic rates and national immunisation was discontinued.

There was a personal tragedy for one of my colleagues among the Siena scientists. Early in 2005, Jeannette Adu-Bobie flew to New Zealand where her laboratory expertise was needed to investigate some anomalies in the assays on immune responses to the OMV vaccine. She had barely started her

---

[p] This and the MenC vaccine (see Chapter 16) in the UK broke new ground in vaccine development and the speed of delivery, precedents that were crucial to the roll out of MenA conjugate vaccine in Africa and to accelerated implementation of Ebola and COVID-19 vaccines (see Epilogue).

work when she became suddenly and seriously ill. In a cruel irony, she had come down with meningococcal septicaemia through exposure in the laboratory. For days, her life hung in the balance as she battled the disease on which she was an expert. Her arms and legs had to be amputated, but she survived. Her courage and irrepressible strength of mind were an inspiration. After a gruelling but triumphant convalescence, she returned to work. Later Jeannette would enrol in the Imperial College Business School as an Executive MBA student and be a finalist in the Student of the Year Award. Miraculously, her career back on track, she would become the project lead and coordinator for the Pathogen Database project in collaboration with the Sanger Centre in Cambridge, UK.

References

[1] Horton, R. Commentary on an octet on vaccines. *Lancet*, 1997, 350, 1192.

[2] Moxon, E.R., Reche, P.A. and Rappuoli, R. *Editorial: Reverse Vaccinology*, 2019, 10:5–6.

[3] Kingsbury, D.T. Bioinformatics in drug discovery. *Drug Development Research*, 1997, 41:120–128.

[4] Watson, J.D. *A Passion for DNA: Genes, Genomes and Society.* Oxford University Press, 2000. ISBN 0 19 850697 X (Hbk).

Chapter

21

A New Era in Meningitis Vaccines

The success of the MenB vaccine in fighting the appalling New Zealand (NZ) meningitis outbreak (1991–2007) was a huge feather in the cap of Chiron Vaccines. It was also the first step in the plan to develop a universal vaccine, one that would protect against all MenB bacteria, not just the variant that had caused the NZ outbreak. I recall vividly a brainstorming session that Rino Rappuoli organised to set out the strategy and milestones. His relentless energy and brilliance energised the meeting; he took on board all the good ideas while diplomatically steering away from those that were not helpful. He knew exactly what outcome he wanted. It was a master class. The NZ outer membrane vesicle vaccine protected against only 20% of the MenB bacteria that caused meningitis. From the genome sequence, there were data on 28 promising vaccine antigens from which to close the gap on the remaining 80% of isolates. None on their own was sufficient to develop a universal vaccine, but a combination of three genome-derived vaccine proteins, together with the NZ vaccine, gave the best breadth of coverage (see Figure 21.1). This became the basis of the Chiron 4 Component MenB vaccine, *4CMenB*.

A few months later, there was an unexpected turn of events. At an international meeting in Oslo in 2002, scientists from the vaccine manufacturer, Wyeth,[a] announced the discovery of an important protein that would protect against the majority of MenB bacteria. Had we missed this important protein in our genome analysis? It turned out that the amino acid sequence of the Wyeth protein was identical to one of the three Chiron genome-derived

---

[a] Wyeth, later acquired by Pfizer, had originally been called Wyeth–Lederle, acquired from Lederle–Praxis which had been the first to license a *Hi*-b conjugate vaccine.

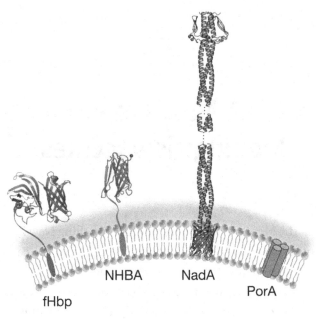

Figure 21.1   Diagram showing three recombinant protein components (vaccine antigens). From left to right these are: factor H binding protein (fHbp), Neisseria Heparin Binding Antigen and Neisseria adhesin A (NadA). These three recombinant proteins and Porin A (PorA, extreme right) from the New Zealand Outer Membrane Vesicle (OMV) vaccine are the four components of the MenB (4CMenB) vaccine that induce protective bactericidal antibodies.

antigens.[b] Wyeth was breathing down our necks and there was urgency to begin clinical trials of the Chiron vaccine as soon as possible.

Before any newly developed vaccine can be given to humans it must first go through rigorous safety testing,[c] investigations that include experiments in animals, a form of biomedical research that is contentious. Although classical laboratory research, such as cell cultures, chemical or physical measurements and even sophisticated computer simulations have their place, many biological functions can only be elucidated by experiments on live animals. The public expects everything reasonable to be done to ensure the safety of vaccines that

---

[b] The independent identification of this protein by Chiron and Wyeth would later result in a hotly disputed series of legal challenges and counter challenges over intellectual property rights and patent infringements.

[c] The guidelines are set out by WHO. The data are reviewed by regulatory agencies, most prominently European Medicines Agency (EMA) [Europe] and Food and Drug Administration (FDA) [USA].

are given to humans. The enormity of the public's condemnation — should a vaccine (or any medicinal product) turn out to be unsafe — is understandable: *primum non nocere*[d] — first, do no harm. It is why, given that the public expect that there will be stringent and rigorous investigations to ensure safety, the regulatory authorities include a requirement for testing vaccines in laboratory animals. In my opinion, there are and have never been convincing philosophical or ethical reasons against animal research. It is an inescapable fact that much of what we know about biology is based on information from animals, including humans. Prohibiting the use of animals for scientific research would logically require cessation of one of the most important sources of knowledge that underpin our understanding of biology. But opponents of animal research act as though no harmful consequences could result from abandoning this research, despite the scrupulous legislation to ensure that the use of laboratory animals for medical research is conducted in a manner that minimises their misuse or suffering.

As I cycle to work each day, I pass a large building[e] dedicated to biological experiments on animals in South Parks Road, Oxford, where on many days of the week, there are a scattering of police casting a watchful eye on protesters displaying placards and billboards calling for a ban on the use of animal experiments. How many of them and their children, I wondered, were immunised with the many vaccines that protect against infections that include smallpox, yellow fever, tetanus (lockjaw), polio, diphtheria, meningitis and many other deadly pathogens? Yet, few would accept being given a new untested vaccine without a substantial measure of reassurance that it would not cause them harm. As I write, we are in the middle of a pandemic. Animal studies have been an essential part of the research and development of the vaccines against COVID-19 without which our lives and the well-being of mankind all over the world would be entirely and drastically different. It's important but very challenging to strike the right balance between sensible, necessary restrictions on animal testing and yet do all that is reasonable to develop vaccines

---

[d] The so-called Hippocratic oath. The coining of this term is arguably attributable to Auguste Chomel, a medical pathologist whose teaching in the first half of the nineteenth century warned against the aggressiveness of unproven medical treatments and promoted the benefits of the healing capacities of natural processes.

[e] Oxford Biomedical Sciences Building.

(or the other medicines that so many people take for granted) that are safe and effective in humans.

By 2002, the portfolio of WHO recommended pre-clinical testing on the Chiron vaccine had been completed and submitted to the regulatory agencies in Europe and USA. The evaluation process did not move forward rapidly; month after month there was a plethora of queries and comments that required detailed responses and often more research. The main sticking point was a predictable one. To give the go-ahead to commence clinical trials in infants, there had to be clear evidence of not just the safety but the protective efficacy of the vaccine. Laboratory evaluation of protection depends largely on the bactericidal assay, which measures the ability of sera from immunised animals to kill meningococci. But, given their enormous diversity, the problem was how to reach agreement on which and how many of the MenB bacterial variants should be included in the testing. These discussions with the regulators and their panel of advisory scientists seemed an unending and excessively bureaucratic process, most of which had to be conducted through exchange of formal documents. To be fair, 4CMenB was a complex vaccine and the first containing antigens based on genome sequencing.

Meantime, while all this wrangling with the regulators was running its course, I had my own struggles in Oxford. For years I had been trying to raise funds to build a dedicated vaccine centre on the medical school campus. The Oxford Vaccine Group (OVG) had expanded hugely and the limited facilities within the Department of Paediatrics were completely inadequate. It was not just the lack of space. In just a few years, the stringency of regulations for conducting clinical trials of vaccines had changed dramatically to the point that we were in danger of failing to comply with the required directives.[f,1] An application for National Health Service (NHS) funds through the Oxford Region was accepted but then, to my immense frustration, withdrawn when the government announced a freeze on all NHS Research Funding. Then in 2002 came an opportunity through the UK Government Joint Infrastructure Fund, through which an award to Oxford University provided £6 million towards the capital costs of the *Centre for Clinical Vaccinology and Tropical Medicine*. This new facility was crucial for OVG to go ahead with clinical trials of the Chiron MenB vaccine.

---

[f] Clinical Trial Regulation: European Medicines Agency. Introduced in 2001, set out the administrative provisions governing clinical trials in Europe (see Ref. 1).

For many years, it had been clear to scientists and public health professionals that there was no realistic possibility of doing a MenB vaccine trial that was based on comparing rates of meningitis in immunised and non-immunised children. Although devastating, meningococcal meningitis (or sepsis) occurs in less than one in every 1000 children per year in the UK. The statisticians had done their sums and concluded that, even if every child born in the UK in one year were to be enrolled, the numbers would not give a clear answer. The logistics and cost of attempting a trial of hundreds of thousands of immunised individuals over several years was simply not justifiable. In a meeting held in Washington DC, international experts agreed that the only practical way forward was to compare *bactericidal activity* in blood samples from immunised and non-immunised individuals. This was what had been done in implementing the meningococcal C (MenC) vaccine in 1999 (Chapter 16) and the New Zealand OMV vaccine (Chapter 20). This had provided strong validity to show the correlation between laboratory tests on blood samples and protection against meningitis. The additional complication was that the vaccine comprised four different vaccine proteins, each of which had to be evaluated. It was a far more challenging version of the Professor's question at my inaugural lecture in 1985 (Chapter 8) that had first set me thinking about how variations in the bacterial surface can compromise vaccine effectiveness. Bacterial variation is the "elephant in the room" when it comes to vaccines and none of us wanted to countenance the possibility that the diversity of MenB bacteria might prove too great a challenge. Since it's a general phenomenon[g] — no less true for the other meningitis vaccines — why was it so especially tricky for the MenB vaccine?

The key lies in the different biological characteristics of polysaccharides and proteins. For the conjugate vaccines made from the "sugar coatings" on the bacterial surface, each distinct capsular polysaccharide consists of a chain of conserved sugars that retain an identical structure. The antibodies induced by immunisation recognise this spatial configuration, bind to it and eliminate

---

[g] One of the most familiar examples of microbial variation is the influenza ("flu") virus vaccine that each year is reformulated to adjust to changes in the "coat" of the virus. The vaccine antigens (neuraminidase and haemagglutinin) alter over time necessitating changes to the vaccine to maintain its effectiveness in protecting against the viral variants that are circulating around the globe. Variations in the COVID-19 virus (for example, so-called UK and South African variants) have also threatened to diminish the effectiveness of many of the current vaccines.

the bacterium. Bacterial surface proteins are more variable and therefore trickier. Proteins consist of hundreds of amino acids linked together like a string of beads whose structure resembles a tangled necklace. Just a single change in one of the amino acids, one bead of the necklace, may be enough to alter the shape of the tangle. Antibodies are precise and if they are to eliminate the bacterium, the spatial configuration of the protein must be retained. The problem is that naturally occurring mutations change the amino acid sequence of proteins and these alterations may alter its spatial shape and prevent the antibody from binding. Of course, it would have been so much easier to make a MenB vaccine using the B polysaccharide, but, for the reasons explained in Chapter 15, this was not possible.

Despite all the difficulties posed by the variability of the MenB vaccine proteins, a road map was agreed by the European Medicines Agency and the required data were completed and approved by 2005.[h] After preliminary trials in a small number of adults, extensive investigations in children began in 2006, a couple of years after Chiron Vaccines had been acquired by the much larger pharmaceutical company, Novartis. This corporate shake-up transformed not only the working policies for the underpinning basic science, but also the strategic and operational approach to the clinical trials. Novartis, exploiting its depth of clinical trials expertise, wisely introduced an accelerated programme to expedite results on infants, the age group most at risk of disease but in whom immune responses were expected to be the least strong. It was in these babies that the results were most critical for getting regulatory approval (licensure) of the vaccine.

At a Scientific Advisory Board (SAB) meeting held in California in 2007, there was a lively discussion on the best methods to characterise the MenB bacteria from cases of meningitis. The idea had been to obtain the gene sequences of the four vaccine antigens for each disease isolate. But this information, visualised in the form of "gene trees," was so complicated that even the experts of the SAB (including a Nobel Prize winner) were baffled. If these experts could not get their heads around the data, what prospect was there of communicating it to the public health personnel and government ministers? Something much more user-friendly was needed.

---

[h] The US Food and Drug Administration needed even more time to reach a consensus.

That night, my sleep pattern severely disturbed after the transatlantic flight, I found myself wide awake at 2 am in the morning, thinking about the question. We needed a scheme that assigned a vaccine-type to each of the myriad of meningococcal bacterial isolates. Given that there were four vaccine antigens, the mathematical possibilities came to 16 possible types. I climbed out of bed, opened my lap-top and set out the possible combinations in tabular form. An email to my colleagues on the Advisory Board arrived well in time for them to see it before the morning session. The proposed method of vaccine typing went down well and became the basis of the *Meningococcal Antigen Typing Scheme*. Jet lag sometimes has its advantages.

From 2006 until 2012, clinical trials in more than 7000 children, toddlers and infants[2,3] were carried out to evaluate the safety, immune responses and protective efficacy of 4CMenB. The vaccine was well tolerated although it caused mild fever in about 18% of babies in the first two days following immunisation, the episodes were transient and were lessened by giving paracetamol.[i,4] However, the million-dollar question was whether the vaccine worked. Using the bactericidal assay, the "gold standard" correlate of protection, at least two-thirds of the immunised children had protective antibody responses, a brilliant result that vindicated the genomic approach.[5] The date was March 2009 and it was necessary to assemble all the results and prepare a dossier for the European Medicines Agency (EMA), an undertaking that was completed by June of 2010. Now came the long wait and anxiety of whether the agency would approve the vaccine.

Figure 21.2   Mariagrazia Pizza.

On November 16, 2012, as I was travelling from Geneva to Siena, I received a call on my mobile phone. It was Mariagrazia Pizza (Figure 21.2), the MenB project leader, the pitch of her voice resonating with excitement and emotion.

---

[i] The occurrence of fever did increase the risk of hospital admissions in the three days after immunisation (see Ref. 4).

The connection was so poor that I needed several iterations to be sure that I had not misheard her news: the EMA had approved the 4CMenB vaccine. It was an incredible moment, the culmination of more than 15 years of research that had been kick-started by the sequencing of the *H. influenzae* genome back in 1995. I was taken completely by surprise since I had not expected a decision from the EMA for several more months. I arrived in Siena just in time for the celebrations. It was highly emotional; some of the key scientists were in tears from joy and relief over what had been achieved. A special meeting in the main lecture theatre had been hastily arranged to relay the breaking news to all Novartis scientists in Siena.

Within weeks, 4CMenB was officially licensed for use in Europe, Australia and Canada. In 2013, almost all cases of bacterial meningitis in the UK were caused by MenB because *Hi*-b, MenC and pneumococcal conjugate vaccines had been so successful. Now, at long last, there was a MenB vaccine. But, because it had not been possible to conduct large comprehensive studies prior to its implementation, its efficacy in the "real world" was unknown. Working with an Oxford colleague, Martin Maiden, we were able to predict the likely impact of the 4CMenB vaccine in the UK, because Christoph Tang, a former member of my research group, had obtained complete genome sequences of all the MenB isolates from cases of meningitis that had occurred in the UK since 2011. By matching the sequences of invasive disease isolates to the vaccine antigens of 4CMenB, we calculated that at least two-thirds of the cases were preventable. The UK was uniquely positioned to provide the required "real time," nationwide data on the effectiveness of the vaccine. But, given the uncertainties[j] and high cost of the vaccine, would the UK government recommend its introduction into the UK routine immunisation programme? In 2013, almost all cases of meningitis in the UK were caused by MenB because *Hi*-b, MenC and pneumococcal conjugate vaccines had been so successful. Now, at long last, there was a vaccine predicted to be effective against most cases of MenB.

The UK government turned to its advisory body, the Joint Committee on Vaccines and Immunisation (JCVI). The JCVI deliberations took place

---

[j] Although direct protection of 4CMenB was estimated (based on bactericidal activity) to be ~73%, there were at the time no data on herd immunity.

from 2011 to 2013,[k] subject to the strict guidelines set out by the National Institute for Clinical Excellence (NICE). The National Health Service requires that all recommendations for proposed interventions, including vaccines, be evaluated for their *cost-effectiveness*, a metric to compare the financial savings to the nation. These same criteria are used whether it's, for example, a hip replacement, dialysis for kidney failure, a course of cancer treatment or a routine immunisation. The key principle involved is for the NHS to allocate its limited health budget using a rational, economic basis. Often referred to as the "level playing field," the validity of cost-effectiveness studies is fraught with ethical and fiscal complexities on which economists are sharply divided.

The difficulties facing the JCVI experts in carrying out the cost-effectiveness analyses on 4CMenB involved many uncertainties, including selecting the most plausible figure to represent the yearly number of cases of meningococcal disease, given the historic variations in incidence. As discussed previously, there had been no clinical trials to show that the vaccine was protective, only data showing that 4CMenB induced antibodies that were predicted to prevent disease. Crucially, the potential of the vaccine to reduce the spread of meningococci (herd immunity) and thus reduce additional cases of sepsis and meningitis was unknown. This was important because community, rather than individual, protection had been shown to make a substantial contribution to the success of the previously licensed meningitis vaccines.

In July 2013, the JCVI published its interim statement. It was a shock announcement. The mathematical models concluded that 4CMenB would not be cost-effective in the UK infant vaccination programme, irrespective of its purchase price. The impact of the JCVI's deliberations was seismic. If their conclusion was accepted, implementation of a vaccine that had taken almost two decades of costly research was a non-starter and — worst case scenario — might be shelved indefinitely. Although the JCVI had followed, to the letter, the guidelines of NICE's framework for making their decision, it sent an ominous signal to researchers and pharmaceutical companies, one that risked discouraging future commitment to the research and development of new vaccines. After all, for many years, public health experts had been advocates

---

[k] The sub-committee met a total of five times (February 2011, July 2012, January 2013, April 2013 and September 2013).

of the need for a vaccine against MenB. Now the JCVI seemed to have pulled the carpet from underneath them. There was consternation and scepticism.

The credibility of the JCVI's conclusions was especially controversial given that a peer-reviewed analysis[1,6] in a reputable journal had shown that the vaccine could be cost-effective if purchased at a low enough price. It quickly became clear that the reason for this discrepancy depended on differences in the information used in the JCVI modelling, not mistakes in the cost-effectiveness (mathematical) calculations. This made sense, because differences in input data, such as hospital costs and long-term care for survivors of meningitis with serious disabilities, were known to have a major impact on the calculations. For example, one metric used by government health economists resulted in a value placed on a child's life of just 27 years, meaning the benefits of saving a child who lived for another 60 years (and the long-term care costs if a victim is left badly disabled) were inadequately accounted for given that most cases of bacterial meningitis occur in children aged less than 5 years. Neither did it account for the effect of the disease on parents' lives and NHS litigation costs. These amounted to £28 million compensation that went to families of children left permanently disabled by meningitis missed by GPs between 2008 and 2012.

But the reaction of so many, especially families who had experienced the devastating impact of death or disabilities from meningitis, was a sense of injustice and outrage. To them, the flaws in cost-effectiveness analyses came down to something simple; it measured health in monetary terms with little or no consideration of the humanitarian costs. What price ought to be placed on the sudden death or lifelong disability of a previously healthy baby? What about the importance of allaying fear and anxiety through provision of an effective vaccine? Widespread educational leaflets in the UK were at pains to emphasise that, in the early stages, it is often not possible to distinguish a baby with a self-limiting, benign viral infection and one whose illness will turn out to be meningitis. It's why fever in a baby is so worrying; failing to spot meningitis in its early stages is one of a doctor's worst nightmares.

---

[1] Ironically carried out by the same research team that had been commissioned by the JCVI (see Ref. 6).

The interim JCVI recommendation was considered by many to be a betrayal of public interest and advocates of the 4CMenB vaccine quickly mobilised themselves to voice their dissent. The Meningitis Research Foundation, a prominent UK Charity, compiled a several-hundred-page rebuttal. Editorials appeared in the *British Medical Journal* and *The Lancet*, the latter venturing:

"There seems little doubt that the public will react strongly to the continuing deaths and disabilities from meningococcus B that will occur in the absence of immunisation. ... Advocates of this and other vaccines will not and should not remain silent."

Fortunately, the JCVI had a "Get Out of Jail Free" card.[m] It was obliged to seek input on the interim position statement from all interested parties. In fact, many public health experts supported the JCVI's position, arguing that a rare disease, however serious, might not be the best use of limited NHS funds. A spokesman for the Royal College of Paediatricians and Child Health warned that: "money spent on the 4CMenB vaccine is money not spent on something else." An international expert on meningococcal disease told me that he was ambivalent given that the vaccine was less than 80% effective and there had been a sharp decline in the incidence of MenB disease in the UK. But others argued vehemently on behalf of the vaccine. More than 100 medical scientists, practising physicians and nurses signed a letter to *The Times* newspaper deploring the decision:

"I have worked with children desperately ill from meningitis and septicaemia my entire working life and like everyone fighting these diseases, I know just how difficult it is to diagnose and treat. The only way to deal with it effectively is through vaccination."

wrote the lead author of the letter, the head of the intensive care unit of a leading hospital and medical school in London. This emotive plea resonated with me as I recalled my own personal experience. My own patient, Julia, had died

---

[m] A part of the board game Monopoly, which is a popular metaphor for something that will get one out of an undesired situation.

within hours[n] and had been such a strong motivation for my research efforts over more than two decades. Sir William Osler's adage came to my mind:

"Medicine is learned by the bedside and not in the classroom … see and then research … But see first."

The JCVI asked for the cost-effectiveness analysis to be re-done using revised and updated data. This time, the mathematical model showed that the vaccine could be cost-effective if given to infants as part of the routine immunisation programme at a low but undisclosed price. The UK used these data to indicate to Novartis the price that could be afforded.[o] Even with a substantial discount, the company stood to benefit from the purchase of millions of doses of 4CMenB. More importantly, it would send a positive signal to other countries that the UK had confidence in their vaccine. Of huge importance, national use of the vaccine would provide an opportunity for post-implementation surveillance to provide data on vaccine effectiveness in the real world — as distinct from estimates based on laboratory tests.

The press had a field day, asserting that there had been a U-turn. The government was forced onto the back foot. According to law, the revised JCVI recommendation obliged the Treasury to find the funds to purchase and implement the vaccine, at a time when the government had put in place an aggressive policy of austerity.[p] The political manoeuvring became aggressive. It was alleged that the ministry of health had deliberately leaked a confidential letter to the Chief Executive Officer of Novartis accusing the company of proposing an unrealistic price for 4CMenB. Nonetheless, a discounted price was eventually agreed and in September 2015, based on the recommendations of the JCVI, 4CMenB vaccine was introduced into the UK infant routine programme to be given at 2, 4 and 12 months. Here at last was the ideal opportunity to find

---

[n] See Prologue, p. xx.

[o] The stand-off between the UK government and Novartis lasted several months and was not resolved until GSK's acquisition of Novartis (March 2015) resulted in a lowering of the price of the vaccine. Details remain confidential, but reports suggested that the final price per dose was around £20, less than a third of its list price.

[p] In the UK, between 2010 and 2019, more than 30 billion in spending reductions were made to welfare payments, housing subsidies and social services by the Conservative Party government.

out the true impact of 4CMenB in preventing meningitis, although it was also obvious that this would need several years of data before any firm conclusions could be drawn.

In the Spring of 2013, worried staff at Princeton University, USA asked the eight thousand students to stop kissing. The campus had been hit by an outbreak of meningococcal meningitis. A young woman from a nearby college died from the disease within days of attending a party with the Princeton football team. People close to the victims panicked and took antibiotics hoping to hold the disease at bay. Anxious Princeton administrators appealed to the US Federal Government for a waiver that would allow them to buy and administer 4CMenB, whose approval in the United States (US) was still stalled in the onerous and lengthy bureaucratic approval process of the FDA. The US Center for Disease Control intervened on Princeton's behalf, and the FDA gave permission to immunise students with 4CMenB. No more cases of the disease were reported, but such observational data did not rule out that perhaps this would have been the case if no vaccine had been given. Nine months later, cases of MenB meningitis showed up at the University of California, Santa Barbara. No one died, but one victim had to have both feet amputated. Once again, a federal waiver paved the way for students to be immunised with 4CMenB. Meanwhile, in Canada, 4CMenB was given to more than 43,000 individuals in Quebec because of an outbreak of meningitis. Although no further cases occurred, the numbers were too small to know whether the vaccine had had an impact, although two cases occurred in unvaccinated persons and the disease continued unabated in surrounding regions. Only through years of observation in a population the size of the UK could accurate information on the vaccine's effectiveness be obtained — and not all was going smoothly.

On Valentine's Day, February 2016, a 2-year-old girl was admitted to Accident and Emergency with a rash on her forehead.[7] She died days later from MenB meningitis and septicaemia. Because of her age (she had been older than 5 months at the time of the vaccine introduction), she had been ineligible to receive the 4CMenB vaccine. Her death sparked a remarkable public campaign to make the vaccine routinely available not just to infants, but also older children. The clamour for this change in policy was heightened when the international rugby star, Matt Dawson, one of the heroes of England's victorious 2003 World Cup victory, entered the fray. His own child had survived meningococcal meningitis, but only after he and his wife had experienced what they described as "two weeks of hell." More than 800,000 signatures were obtained for the most popular petition in parliamentary history, debated in the House of Commons in March 2016. The government rejected the petition, although cynically delaying its response until February 2018, which read as follows:

> "The NHS budget is a finite resource. Offering the vaccine outside of JCVI's advice would not be cost-effective and would not therefore represent a good use of NHS resources which should be used to benefit the health and care of the most people possible."

By June 2016, 500,000 infants had received two out of the three doses. Public Health England reported that the effectiveness of 4CMenB was 83%.[8] But a close inspection of these findings showed that even with these relatively large numbers, this estimate needed to be interpreted with caution. The range of possible values, the so-called confidence interval, ranged from 24% to 95%. But, looked at in a different way, instead of the 74 expected cases, there had been only 37. The vaccine had halved the expected number of meningitis cases in young babies. By January 2020, the same scientists reported that there had been 63 cases instead of the expected number of 274. An accompanying editorial in the *New England Journal of Medicine*[9] concluded that although this was good news, an improved vaccine was needed. There was a further downside; a study in Southern Australia showed that 4CMenB had no impact on reducing person-to-person spread of MenB. In contrast, the reduced exposure through community protection (herd immunity) using conjugate vaccines had accounted for more than 50% of the reduction in cases of meningococcal meningitis. After more than 20 years of research on the 4CMenB vaccine, the vaccine had achieved notable success, but improving the breadth of coverage and, above all, the imperative of inducing community protection[9,10] remain major challenges for future research.

---

[q] The social distancing policies, introduced to contain person-to-person spread during the COVID-19 pandemic, have had a drastic impact on the number of cases of meningitis. An international consortium of scientists from 26 countries (see Ref. 10) have recently documented this profound change in epidemiology. In the UK alone, data from a three month period (April to June 2020) showed a 76% reduction in cases of meningitis; from 121 cases in 2019 to 29 in

References

[1] Directive 2001/20/EC of the European Parliament and of the Council 4 April 2001 https://ec.europa.eu/health/sites/health/files/files/eudra.

[2] Su, E.L. and Snape, M.D. A combination recombinant protein and outer membrane vesicle vaccine against serogroup B meningococcal disease. *Expert Review of Vaccines*, 2011, 10:575–578.

[3] Vesikari, T. *et al.* Immunogenicity and safety of an investigational multicomponent, recombinant, meningococcal serogroup B vaccine (4CMenB) administered concomitantly with routine infant and child vaccinations: Results of two randomised trials. *Lancet*, 2013, 381:825–835.

[4] Kapur, S. *et al.* Emergency department attendance following 4-component meningococcal B vaccination in infants. *Archives of Disease in Childhood*, 2017, 102:899–901.

[5] Donnelly, J. *et al.* Qualitative and quantitative assessment of meningococcal antigens to evaluate the potential strain coverage of protein-based vaccines. *PNAS*, 2010, 107:19490–19495.

[6] Christensen, H. *et al.* Introducing vaccination against serogroup B meningococcal disease: an economic and mathematical modelling study of potential impact. *Vaccine*, 2013, 31:2638–2646.

[7] https://www.dailymail.co.uk/health/fb-5595427/WHO-FAYE-BURDETT-TODDLER-DIED-MENINGITIS-B.ht.

[8] Ladhani, S.N. *et al. New England Journal of Medicine*, 2020, 382:309–317.

[9] Harrison, L.H. and Stephens, D.S. Good News and Bad News — 4CMenB Vaccine for Group B *Neisseria meningitidis*. *N. Engl J Med*. 2020, 382:376–378.

[10] Brueggerman, A.B. *et al.* The Invasive Respiratory Infection Surveillance (IRIS) Initiative reveals significant reductions in invasive bacterial infections during the COVID-19 pandemic. https://doi.org/10.1101/2020.11.18.20225029.

2020. It's a powerful reminder of the importance of person-to-person spread of pathogens and why the herd immunity (indirect protection) resulting from vaccines is so important.

Chapter

22

The Most Important Medical Intervention in History

The development of vaccines against some of the major causes of bacterial meningitis has radically transformed the global picture, although the fight is most certainly not over. But, as I ventured at the beginning of the book, vaccines are the greatest success story of modern medicine and it's hard to imagine a world without them. Before them, the chance of dying from small-pox was 30% and the lives of survivors were blighted by tell-tale, disfiguring facial pock marks. Look at a dollar bill with its picture of George Washington; the face you see is very different from what people who knew him would recall. In real life, the first American president's face was pitted and scarred, although the unsightly consequences of smallpox were not always a social disadvantage. Advertisements for servants in the eighteenth century often requested that applicants be pock-marked to ensure they'd been infected, were immune, and therefore couldn't catch and spread the virus within the household. Thanks to immunisation, smallpox was eliminated from the planet by 1977, the only human disease to have been completely eradicated, but not before it had caused the deaths of hundreds of millions of people.

Readers under the age of 60 may find it hard to imagine the widespread anxiety caused by outbreaks of polio. Fear of getting the virus caused people to flee from cities and towns in their thousands; meetings and public gather-ings were virtually abandoned. I remember as a child being warned not to buy ice cream, to stay away from swimming pools, and I was forbidden to go to the cinema. Epidemics of polio literally terrorised communities whose lives were transformed from the norm into one of panic. At the peak of a polio epidemic, all available hospital beds were occupied, the wards filled with paralysed people, many of whose chest and diaphragm muscles were so weak they couldn't breathe for themselves. Their lives depended on the availability of artificial respiration using the infamous tank respirator or iron lung. Imagine the terror of not being able to breathe because your lung muscles are paralysed so that you are gasping for air. The medical team puts you into something that looks like a metal coffin, sealing you in; there's a strange sound, like a giant bellows, and suddenly you can breathe. But for many, life in an iron lung was one of intolerable boredom and many did not recover. Today, universal immunisation has resulted in the near global elimination of polio from all but a handful of countries.[a]

Figure 22.1   The first iron lung was used at Boston Children's Hospital on October 12, 1928, to save the life of an 8-year-old girl. It was pioneered by Philip Drinker, Louis Agassiz Shaw and James Wilson of the Harvard School of Public Health.

At the top of the beautiful Victorian pinewood stairs of the Sir William Dunn School of Pathology in Oxford (the very same building where Howard Florey and Ernest Chain began their pioneering work on penicillin in 1939)

hangs a painting which illustrates powerfully an age without mass vaccination. A young child lies on a makeshift bed, sleeping or perhaps comatose. Around the cot are discarded, blood-soaked wipes, possibly caused by the diphtheria bacterium. At her side sits a thoughtful physician, caught in the Rodin-like pose of *Le Penseur* and bearing an uncanny resemblance to the acclaimed scientist and microbiologist, Louis Pasteur.

Figure 22.2   *The Doctor* (1891) by Sir Luke Fildes (1843–1927). Tate Gallery, London.

The frustration of the doctor, unable to do anything to save the child, is etched in his reflective but despairing features. In the background, the exhausted mother slumps over the table, and the stoic, facially expressionless father awaits the inevitability of the death of their daughter. Today, routine immunisation with the diphtheria vaccine has virtually eliminated the disease in all parts of the globe. The occasional cases that do still occur are in communities where

---

[a] The poliovirus is now only endemic to three countries in the world — Afghanistan, Nigeria and Pakistan. In 2020, there were 105 cases.

societal breakdown or other disasters have interfered with routine immunisation programmes.

In rural provinces of South Africa, the birth of a Zulu child is celebrated in an ancient ceremony that involves the smearing of animal faeces on the umbilical region to ward off evil spirits. Among the myriad of bacteria in the dung are spore-forming bacteria called *Clostridium tetani*, the cause of a dreaded disease that, through a toxin, causes life-threatening muscle spasms. Tetanus of the newborn is a catastrophe. Imagine what it must be like for the joyful but exhausted mother after giving birth to see her infant convulsed with uncontrollable, atrociously painful muscle contractions — the same disease that in older people is called lockjaw. Yet, if women are immunised before giving birth, their immunity is transferred to the unborn infant and neonatal tetanus is completely prevented.

As I discuss in the next chapter, it is hard for me to understand why anyone would not want to do something that benefits their own and other's health and that of young children for whom they have responsibility. Although confidence in immunisation remains high in most parts of the world, opposition to it cannot be ignored; it ranges from hesitancy to the extreme views of activists, often called *anti-vaxxers*, whose social media campaigns of emotive language and images reach hundreds of thousands of people in an attempt to engage people's worst fears (see Chapter 23).

Ironically, I must point out that the medical profession itself was responsible for a breakdown in confidence in two of our most important vaccines; whooping cough (pertussis) and measles. Babies who are unfortunate enough to get infected with pertussis in the first six months have a 50% death rate. In the 1930s, a vaccine, made from bacterial cells (and therefore called the whole-cell pertussis vaccine) was developed and, after some improvements, was shown in the 1950s to be about 80% effective. Through the World Health Organisation's expanded programme in immunisation, over 80% of all infants receive pertussis vaccine,[b] an intervention that prevents around 750,000 each year worldwide. It was well known that, after being given the whole-cell pertussis vaccine, a few babies became unwell with fever, irritability, refusing feeds

---

[b] Given as a combined vaccine to prevent diphtheria, tetanus and pertussis (DPT). In 2004, an alternative vaccine became available as an alternative to the whole cell vaccine. Although causing fewer side effects, it is less effective in preventing spread of pertussis bacteria.

and, of particular concern, prone to high-pitched inconsolable screaming. In 1974, paediatricians from the UK's Great Ormond Street Hospital published a report suggesting that the vaccine caused brain damage known as pertussis *encephalopathy*. *The Daily Mail* published an editorial raising fears that "there are possibly hundreds of teenagers with the body of an adult and the mind of a child because they were vaccinated." It prompted questions in parliament and huge anxiety among the medical profession and the general public. Concerns over the vaccine and brain damage spread widely throughout Europe and North America, prompting a lack of confidence in the vaccine. Within a year or two, the uptake of the vaccine in the UK had fallen by around 50%. The furore prompted an investigation, *The National Childhood Encephalopathy Study*, although its conclusions were not published until 1981. All the study could confirm was that giving the vaccine was associated with a higher incidence (1 in about every 110,000 infants) of *febrile seizures*. Almost all of these babies got better quickly, were healthy and developed normally. In a few, the fever unmasked latent epilepsy[c] that would have emerged later. Today, it is known that pertussis vaccine does not cause brain damage.[d,1] But, the decline in immunisation caused three major epidemics of whooping cough in the UK. The first started about 1977 and accounted for 102,500 cases of pertussis, thousands of hospital admissions and 36 deaths.[2]

Then there was Andrew Wakefield, an apparently eloquent and charismatic British paediatric gastroenterologist, who had a long-standing interest in measles. In 1993, he published an article wrongly suggesting that the measles virus (naturally occurring or the weakened form in the vaccine) was the cause of an inflammatory bowel disease, called Crohn's. In 1998, at a sensational press conference, he announced that measles virus in the vaccine caused *autism*. Based on investigations in 12 children and published in *The Lancet*, the findings were subsequently shown to be fabricated. Worse, Wakefield was paid more than half a million pounds as a scientific advisor to a law firm seeking indemnity for brain damage allegedly caused by the combined measles, mumps and rubella vaccine, known as MMR.

---

[c] The prevalence of epilepsy in the population is around 3%.

[d] In 1982, an incendiary and fraudulent US documentary stated as fact that the vaccine caused brain damage (see Ref. 1).

Media coverage of this scandal left people confused, unable to know whether the claim was true or false; MMR vaccination levels fell from over 90% to 80%, resulting in outbreaks of measles[e]. It took years before in-depth investigations by a journalist[3] and the General Medical Council exposed the extent of the fraud. Wakefield was struck off the Medical Register, *The Lancet* belatedly retracted the article and the disgraced doctor emigrated to the United States where he has become a celebrity, befriended and supported by ex-President Trump and a coterie of anti-vaxxers. Unrepentant and a disgrace to our profession, he still claims that measles vaccine, when given together with mumps and rubella vaccines, causes autism.

Measles is one of the world's most contagious diseases and is preventable by immunisation. Although often considered one of those diseases that everyone used to get as a child[f] (the cause of just a few days of a nasty illness with high fever), this is highly misleading. Measles is most certainly not benign and is the cause of many serious complications.[g] One of the most serious complications of measles is the form of Brain Fever known as encephalitis.[h] Like meningitis, encephalitis can be life-threatening. Measles encephalitis can occur during the height of the illness — coinciding with the typical rash when the amount of virus in the blood is at its peak and spreads to the brain. But there is also a delayed form of encephalitis caused by an immune reaction to the virus — not against measles virus in the brain itself, but elsewhere in the body. Days or weeks later, this uncontrolled immune reaction causes severe brain inflammation.[i] It's another reminder of how our host defences — so vital in fighting germs at the beginning of an infection — can be a two-edged sword. Paradoxically, as I mentioned in Chapter 1, it is the ensuing inflammation resulting from our immune responses to germs that causes so much of the tissue damage that we experience as disease. Measles encephalitis

---

[e] Over a four-year period, there were 10,794 cases of measles between (1998–2001).

[f] Prior to the availability of a highly effective vaccine (1963), nearly all children got measles before age 15 years.

[g] These include severe gastroenteritis, ear and eye infections, pneumonia and inflammation of the brain with seizures.

[h] Whereas meningitis refers to inflammation of the linings of the brain (meninges) and the spinal fluid (see Chapter 1), the term encephalitis refers to inflammation of brain tissue itself.

[i] An even more delayed, very rare form of encephalitis, called SSPE (sub-acute sclerosing panencephalitis) also occurs, especially in persons with weakened immune systems.

occurs in one in every thousand cases, but it has never occurred after measles immunisation.[4] So, in the UK alone, the measles vaccine can prevent around 250 cases of measles encephalitis each year (many of which are fatal) as well as other complications such as pneumonia.

An example of delayed inflammation resulting in brain fever, now largely forgotten, occurred worldwide from October 1918 until 1926. It was known as *encephalitis lethargica* and coincided with the notorious Spanish influenza pandemic. The typical findings — fever, headache, impaired vision and profound lethargy — were described in Oliver Sachs's 1973 book.[5] It killed half a million people and although some recovered completely, many survivors developed long-standing neurological problems, the most common being a form of Parkinson's Disease that typically occurred two years or more after the initial illness. After affecting more than a million people across the world, very few cases of encephalitis lethargica occurred from about 1926 onwards. The mysterious decline once the Spanish "flu" had run its course suggested that the virus might have been the cause, but this remains to this day an enigma. There is little evidence for a direct link to the virus, but given the striking epidemiological association with pandemic influenza, many experts believe that the virus potentiated an immune reaction that resulted in this devastating form of encephalitis.[j,6] There are concerns that future influenza pandemics might bring about a recurrence of this devastating sequel, although there have been no reported instances of a similar problem in the three subsequent pandemics.[k] It raises the question of whether influenza vaccines, unavailable of course until the 1940s, may prevent these terrible neurological complications.

The role of immunisation in preventing the devastating complications of post-viral encephalitis is a matter of extreme public health importance right now. Even before the WHO officially classified COVID-19 as a pandemic (March 11, 2020), reports of serious nervous system complications were apparent. A few weeks after the peak of the UK epidemic in March 2020,

---

[j] Von Economo, who first described encephalitis lethargica, proposed that the influenza virus spread to the brain by invading the mucosa within the nose, tracking along the pathway of the (olfactory) nerve that is responsible for our sense of smell. Post-mortem examinations showed brain cells (neurons) tangled in a mesh of protein polymers and a reduction in size of parts of the brain (see Ref. 6).

[k] 1957–1958 (H2N2 Asian "flu") and 1968 (H3N2 Hong Kong) and 2009–2010 (H1N1).

I received an email from a medical colleague who was suffering from the condition often called *chronic fatigue syndrome*. After recuperating from a relatively mild respiratory illness, he found himself battling the nightmare of mental confusion and extreme lethargy. He was incapacitated, a disability experienced by thousands of others who have suffered from a spectrum of neurological problems (emotional instability, psychosis and *brain fog*) — part of the syndrome now called *Long COVID*. The havoc caused by a later-onset, exaggerated immune response to a variety of infections, including COVID-19, is metaphorically called a "*cytokine* storm." Cytokines are one of the many potent inflammatory molecules — chemical messengers — triggered by invading microbes. Early in an infection, inflammation defends our bodies against microbial invasion and is beneficial. But inflammation also causes tissue damage, a trade-off exemplified by the later stage complications of infection with COVID-19 that damages the lungs and many other body tissues including the brain. Treatment with dexamethasone, a potent anti-inflammatory molecule, can be lifesaving against COVID-19 and has proved valuable in the treatment of other serious infections, including meningitis. But if dexamethasone is given too early in the infection, the dampening down of inflammation can allow an invading microbe to multiply and overwhelm the body. It is all about timing and why it is so important to understand the complex biology of our bodies and their interactions with microbes. Of course, pre-empting infection through immunisation is the ideal intervention and there is no more exciting milestone in the history of immunisation than the truly remarkable achievement of vaccines against COVID-19. It's a story in which my own university has played a major role.

It all began December 31, 2019, with an alarming report of a cluster of cases of life-threatening pneumonia from Wuhan, China. Those who had become sick had in common that they had all visited the city's large seafood market and by early January, the causative virus (now known as COVID-19) had been isolated. By January 10, 2020, the genetic sequence of the virus had been teleported across the world,[1] literally at the speed of light. Lessons had

---

[1] Virus sampled from some of the first COVID-19 cases seen in Wuhan was sequenced by the Chinese scientist Zhang Yongzhen in Shanghai. Under pressure from the Chinese government not to make the information public, but realising its global importance, he shared the sequence data with his Australian collaborator, Eddie Holmes. Within an hour of receiving it, Holmes

been learned on pandemic preparedness after the SARS (Severe Acute Respiratory Syndrome) scare of 2003, the 2010 influenza outbreak and the more recent experiences with MERS (Middle Eastern Respiratory Syndrome, 2012). In fact, nothing had been the same concerning pandemic preparedness since 2013,[7] when a new lethal avian influenza virus in China threatened to become a pandemic for which the world was not prepared. In previous pandemics, vaccines had become available only after the pandemic peak, and therefore they were too late to be useful. On this occasion, based on the genetic blueprint of the influenza virus, scientists used the newest technology to synthesise the key vaccine components within hours. Within a week, vaccine seed lots were ready for testing in animals.

Armed with this background of massively improved technological know-how, Oxford scientists had set up a generic technology platform that enabled an accelerated approach to making viral vaccines. When the COVID-19 sequence from Wuhan became public, using recombinant DNA techniques, a team led by Sarah Gilbert inserted the genetic code of the COVID-19 spike protein into a

(a)　　　　　　　(b)

Figure 22.3　Professors (a) Sarah Gilbert and (b) Andrew Pollard of Oxford University.

harmless, specially constructed chimpanzee virus that could be safely given to humans to induce an immune response to the virus.

Oxford's pandemic preparedness crucially included the 30 years of experience of the Oxford Vaccine Group[m] that I had set up in the 1990s (see Chapter 16). After my retirement, the current Director, Andy Pollard, set up a series of ambitious, judiciously planned human clinical trials that began within a few weeks of the start of the pandemic. Equally important, a partnership with "Big Pharma" (AstraZeneca) allowed efficient manufacturing of millions of

---

put the genetic blueprint of this new coronavirus online, so that it was available worldwide to all scientists.

[m]The capacity to carry out human trials of vaccines depends on having a pre-existing infrastructure of expertise and activities. Assembling such a team requires years of investment if it is to be fit for purpose and meet the challenge of rapid deployment when faced with a sudden pandemic crisis.

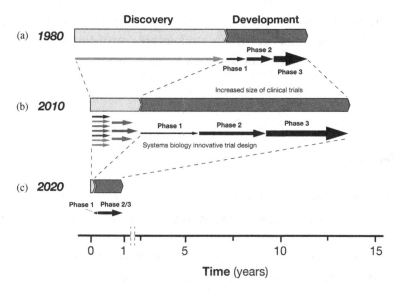

Figure 22.4   Timelines for development of selected vaccines (discovery and clinical trials) represented by: (a) *H. influenzae* type b conjugate vaccines (1974-1987); (b) pneumococcal (7-valent) conjugate vaccines (1983-2000). Contrast these profiles for vaccines against endemic, commensal bacteria that cause meningitis with: (c) accelerated development of vaccines against the pandemic virus COVID-19 (2020). Figure adapted from Ref. 7.

doses of vaccine months before the outcome of clinical trials showed that the vaccine was indeed safe and effective. In less than a year — the starting point being the genetic sequence of the virus on a computer screen! — a vaccine was approved by the Medicine and Healthcare Products Regulatory Agency (MHRA) at the end of December 2020 (see Figure 22.4).

What a triumph for all the scientists of this and the several other COVID-19 vaccines, the most efficient intervention that can control a pandemic that has infected more than a hundred million, killed more than two million and, as emphasised earlier, wrought havoc with the lives of so many, including the impact of Long COVID. As one of the older persons in the UK population, I was happy to be immunised along with hundreds of others who arrived at my wonderful NHS practice to get our jabs in the early part of this year.

Among those prioritised are patients and health workers in care homes, key workers in emergency services and the many who for a variety of reasons are especially vulnerable. How extraordinary it is that, despite the amazing exploits of academic and industry scientists, public health workers and volunteers who made safe and effective vaccines available so quickly, significant numbers of care workers are refusing to be immunised. This is no trivial problem and there are no easy answers — as I discuss in the following chapter.

References

[1] https://www.youtube.com/watch?v=qpUsg4bDH5w DPT Vaccine Roulette 1982.

[2] Baker, J.P. The Pertussis Controversy in Great Britain 1974–1986. *Vaccine*, 2003, 21:4003–4010.

[3] Deer, B. *The Doctor Who Fooled the World: Andrew Wakefield's war on vaccines.* Scribe 2020.

[4] Campbell, H., Andrews, N., Brown, K.E., and Miller, E. Review of the effect of measles vaccination on the epidemiology of SSPE. *International Journal of Epidemiology*, 2007, 36:1334–1348.

[5] Sacks, O. *Awakenings.* Picador 1991.

[6] Von Economo, C. *Encephalitis Lethargica. Its sequelae and treatment,* 1931. Oxford University Press.

[7] Rappuoli, R., De Gregorio, E., and Del Giudice, G. *Proceedings of the National Academy of Sciences,* 118. E2020368118.

Chapter 23

Trust and Mistrust in Vaccines

Given the extraordinary benefits that vaccines have brought to human health, it seems perverse that immunisation is, at least for some, controversial. The problem of vaccine refusal has been identified by the World Health Organisation as one of the top ten threats to global health. The *Global Vaccine Alliance (GAVI)*, an organisation founded in 2000 to facilitate equitable access to vaccines for children from low-income countries, has played a major role in achieving the vaccination of around 760 million children, preventing more than 13 million deaths. Beyond saving lives, vaccines are not just a humanitarian imperative, they are also of profound economic importance and even wealth-creating.[1] So, what has happened to promote the emergence of a widespread and burgeoning mistrust in vaccines? Although concern about the wisdom of getting immunised is not new, it has gathered momentum, especially over the past couple of decades. Interestingly, these fears are more prevalent in socio-economically wealthy countries.[2]

Strong opposition to vaccines and its less extreme manifestation, *vaccine hesitancy*, has become especially prominent when it comes to the measles vaccine; its acceptance has been used as a touchstone of public compliance in vaccine acceptance. Many have attributed refusal of measles vaccine to the rogue British medical scientist, Andrew Wakefield, who, as already mentioned, falsely claimed that it caused autism. However, although a touch paper may have been ignited by Wakefield, it is too facile to pin the wildfire of global decline in measles immunisation uniquely on him. To pursue the metaphor, wildfires happen in a permissive context such as draught, favourable winds and high ambient temperatures, where a small spark can ignite the blaze.

So it was that Wakefield's fraudulent claims lit a fuse that came at a time when there was already an existing decline in measles immunisation leading to global outbreaks in multiple countries literally from A to Z, Australia to Zimbabwe.[a]

There's another issue to consider. Not all children can be immunised against measles for medical reasons. For example, those being treated for cancers or who have diseases that lower their resistance to infections. This is because the measles vaccine is a weakened (attenuated) form of the natural, live virus, so it multiplies in the body. In healthy individuals with normal immune functioning, serious adverse reactions after immunisation are extremely rare. But in those occasional individuals whose immunity is compromised, there is an increased risk of serious side effects from the vaccine.[b] But if most of the population has been immunised, person-to-person spread of the virus is negligible. This so-called "herd immunity" protects others, including those who have not been immunised, from being exposed and getting measles.[c] The snag is that because measles is so highly contagious, herd immunity depends on around 95% of people being immunised to achieve this population-level immunity. This means that almost everyone needs to be immunised, a requirement that can create tensions between the rights of an individual and their obligation to society. To an extent, being immunised can be considered a civic duty. The issue of altruism adds a complicated dimension to the immunisation debate.[3]

Population-wide immunity to measles in the US did reach a point in the year 2000 when there had been no new cases of measles for 12 months and there was talk about whether the virus could be eliminated. Unfortunately, this optimistic scenario was short-lived because there was a drop in vaccine

---

[a] Measles was rare in Australia until in 2019 there was a spike due to active vaccine refusals resulting in susceptible children who became infected largely though imported infections from Thailand. Zimbabwe experienced declining measles immunisation rates from 1996–2004, during which there were more than 500 deaths. The problem was caused by attrition of health care workers, closure of health facilities, breakdown in cold-chain and lack of vehicles and fuel taking vaccines to outreach health services. These two examples illustrate the multifactorial nature of the decline in measles immunisation.

[b] All such people should know that they are at risk and receive expert advice accordingly.

[c] Herd immunity, or indirect protection. has already been discussed in the context of the conjugate vaccines against bacterial meningitis (see Chapter 16). It's a somewhat unfortunate name that conjures up images of rounding up cattle; perhaps community protection is a more appropriate term.

uptake. As a result, there was no longer herd immunity and measles outbreaks were triggered, largely by travel-related importations of the virus that spread rapidly (it's very contagious) to those who were not vaccinated. The US experience was not unique. The same pattern was mirrored elsewhere as several other countries that had virtually eliminated measles had resurgences.[d]

Opposition to immunisation seems to be part of a broader societal malaise of mistrust in governments, institutional experts and other forms of top-down functioning authorities, especially among those who have suffered deprivation and poverty through societal inequality. Disenfranchised and dissenting groups within society have been empowered by social media through which they can articulate a collective voice for their frustrations. Confronted with what are perceived as unpleasant threats over vaccine decisions, social media provides a means (often through misinformation) to counter their feeling of having lost control. Many of the claims are ludicrous, such as the resurrection of a long-discredited idea that the Earth is flat or that there's an underground cabal of secret paedophiles who will imminently assume political power (QAnon). Social media platforms and political consulting companies have exploited these societal fears — often using dubious means to acquire and even pirate personal details — to target vulnerable groups. These expertly tailored messages are highly influential,[e] a means of manipulating people who find relief and belief in alternative ideas that dispel fears and increase a sense of well-being.

In her recent book, *Stuck*,[4] the anthropologist Heidi Larson provides a thoughtful analysis of how trust has been eroded by nationalism and populism. One of her major themes concerns the importance of rumour. "Managing rumours is about understanding and managing the emotions which drive them, not attempting to judge whether they are true or false."[5]

---

[d] Reported cases of measles rose globally by 300% in the first three months of 2019 compared with the same period in 2018. WHO reported 90,000 cases of measles in Europe during the first six months of 2019.

[e] For example, in the influential campaign to discredit measles immunisation in Samoa. Their actions have also been in part responsible for the polarisation exemplified by Brexit in the UK and Trumpism in the US. The messaging is subtle for research shows that it can be counterproductive, reinforcing rejection of trends in those already sceptical. But getting it right can be hugely profitable.

Even the idea that there is a legitimate debate about the value of vaccines is itself unfounded. Vaccines work; to say otherwise is a lie, a malicious peddling of misinformation. Larson[f] concludes that improving trust in vaccines will be an uphill struggle:

> "… today we are in the paradoxical situation of having better vaccine science and more vaccine safety regulations and processes than ever before, but a doubting public."

The current public health attitude to the anti-vaxxers is not to lock horns with them. Confrontation only facilitates their credibility and public profile. Discussions anyway usually prompt further entrenchment than a change of attitude. Rather, efforts are invested in positive education about vaccines and research to understand better the basis of hesitancy. Although the role of social media has radically increased the opportunities to peddle misinformation, fake news and conspiracy theories are often riddled with inconsistencies. One powerful approach is to communicate how to recognise these flaws. Arguably, more sophistication in discerning corrupt social media messaging is needed to counter the torrent of misinformation about vaccines.

But, countering a lack of trust in vaccines is not simple. Although no vaccine is free from possible harmful effects, a part of the challenge is in communicating impartially their benefits and risks. This decision tree is precisely where the reach and power of social media is such an influential means of distorting the pros and cons. Evidence based arguments, no matter how skilful, will not on their own turn the tide and facts alone must be recognised as an inadequate solution. Besides, an important but more mundane impediment is that, in a busy world of competing priorities, missed immunisation visits are as much about fitting everything in than people exhibiting vaccine hesitancy.

So, what can we do? It may be helpful to understand the complexity of what is known to psychologists as the dual process model of behaviour. According to Daniel Kahneman,[6] we respond to the world around us using two different

---

[f] Heidi Larson and her team at the London School of Hygiene and Tropical Medicine have set up the *Vaccine Confidence Project*, a system for early detection of public concerns around immunisation including the measles, mumps and rubella (MMR) vaccine. It has rigorously documented the recent decline in vaccine confidence.

modes of thought and it's worth exploring the concepts that won him a Nobel prize (Economic Sciences) in 2002, detailed in his seminal book *Thinking, Fast and Slow*. One mental process (prosaically called System 1) is the rapid, intuitive response to the world around us. It is an instinctive thought process that we can't switch off. It embodies an evolved, hard-wired mechanism of the human brain through which we make the many snap judgements that are vital to our self-preservation and well-being. We are after all survivors through a Darwinian selection process that favours behaviour that protects our genes (and their heritable legacy to our offspring) in dealing with signs of danger, recognising those who may harm us, avoiding hazards, protecting our loved ones. It's the product of at least 200,000 years of human evolution. In contrast, System 2 thinking (I am using it now as I write) is slow, intentional and demanding; it requires attention and energy. It tires readily and is easily overcome by System 1. But, there's a real downside to System 1. It simplifies, it ignores the rational processes of balancing risks and benefits, thrives on biases (unconscious) and is prey to misinformation. Unfortunately, it's a flaw of human nature that, as Mark Twain observed, allows us to find the "… little bit of truth that enables us to believe things which we know to be untrue."

The beauty of Kahneman's (and his colleague Amos Tversky's) thesis is that it has inspired thousands of experiments that provide compelling evidence that it is integral to our behaviour. There is no escaping System 1 even though it's at the root of many bad decisions, including those that affect our decisions over vaccines. Weighing risks and benefits is a part of our life, but few of us do so in a rigorous and objective fashion. Reliance on ideas, beliefs, spurious rationalisations and feelings of well-being are far more powerful determinants of what we do than a hard-nosed analysis of evidence. Emotions are essential to us in our day-to-day living, and dumping cold explanatory water on them is not an intelligent solution to vaccine hesitancy.

Although there can be little doubt about the importance of science-based, clearly formulated communication to counter the fears and scepticism over vaccines, perhaps it is in the skill and nuancing of this messaging that there is scope for improvement. Overcoming emotional fears, irrespective of whether they are true or false, is dependent on engagement to understand why people are fearful that vaccines may be harmful. We must listen, be respectful and, above all, not patronise. We need to uncover the complexities, the multi-layered and nuanced nature of mistrust in vaccines. A "one-size suits all" approach will fail. The messaging must be tailored according to the cultural context and the shifting dynamics within different societies. We also need a set of modular strategies to manage, improve and sustain confidence in vaccines and this requires a coordinated, comprehensive global initiative. We are interconnected because getting a vaccine affects others through community (herd) immunity.

This is a theme taken up by the writer and literary scholar Eula Bliss.[7] Based on her experiences as a mother, she skilfully explores the inadequacies she feels as life circumstances uncover her unpreparedness to make complex decisions on behalf of her child, a central theme of which is immunisation. But the scope and reach of her text is much broader; she captures the tensions she feels as decision-maker (how much do you need to know and understand to act responsibly?). What she decides on behalf of her children is not just about their welfare. Today, as we face the global challenges of COVID-19 and the certainty that other pandemics will happen in the future, the issue of mistrust in vaccines remains a paramount and unsolved concern.

References

[1] Bloom, D. Canning, D., and Weston, M. The value of vaccination. *World Economics*, 2005, 6:15–39.

[2] A First World Problem. *The Economist.* August 29th 2020.

[3] Steinberg, D. Altruism in medicine: Its definition, nature and dilemmas. *Cambridge Quarterly of Healthcare Ethics*, 2010, 19:249–257.

[4] Larson, H.J. *Stuck. How Vaccine Rumours Start — and Why They Won't Go Away*. Oxford University Press, 2020.

[5] Allport, G.W. and Postman, L.J. The basic psychology of rumour. *Transactions of the New York Academy of Sciences*, 1945, 8:61–81.

[6] Kahneman, D. Thinking Fast and Slow. *Allen Lane*, 2011: 20–30.

[7] Bliss, E. *On Immunity: An Inoculation*. Fitzcarraldo Editions, 2014.

Epilogue

As I look back to the time when I was a medical student and junior doctor in London, I cannot help but reflect on how 1970 — four years after I qualified as a medical doctor — was a turning point in my career. I recall vividly the hours of restless reflection as I gazed from the deck-side railings at the swirling, surging turbulence of the Atlantic waves on my return voyage from South America just before going to Boston. I think *Brain Fever* is an apt metaphor for my mental state; latently energised, I was ready to be challenged.

My exposure to many brilliant clinician-scientists in the US opened the door to research possibilities that I had hardly considered. Until then, in my own eyes, I did not remotely belong to this cadre of academic medics until without warning I was suddenly in their midst and it was expected, even assumed, that I would deliver. There was undoubtedly something very "American" about their (but not my) confidence. How very different the modus operandi of a US residency programme was when compared to the "work experience" that had been characteristic of my prior clinical training. In the UK, the opportunities at that time, let alone expectations, of embarking on serious research were simply not on my radar screen.[a]

Given opportunity and strong support, I found that I had a measure of scientific research creativity which previously I had had few opportunities to assert. The audacity to explore bold, even wild ideas, a lively imagination and belief in oneself are priceless ingredients in doing science. So are discipline and

---

[a] Until the 1980s, funding to support the research careers of UK medical scientists was exceptionally difficult to secure. This all changed when the assets of Burroughs, Wellcome and Company were used to launch an independent charity, the Wellcome Trust. It became one of the largest biomedical funding institutions in the world. It resulted in a marked increase in career awards, fellowships and schemes that had a major impact in supporting medically qualified scientists.

humility, although the amalgam of all these traits may be confusing. Early in my research fellowship in Boston, I recall spending a stimulating hour enthusiastically discussing some ideas for new experiments with a thoughtful and receptive senior colleague. I did most of the talking. The next day, I found a folded hand-written note on my lab bench: *Richard: I think that it's the things you think you know, but don't, that will hurt you.* Touché.

It was my good fortune to arrive in Baltimore just as a new era in biology had begun to make an impact on the medical sciences. Ham Smith, whose research pioneered recombinant DNA and cloning technologies, was my mentor in the late 1970s. Captivated by the sophistry of molecular biology and under his aegis, my research transitioned from classical to molecular microbiology as I set about isolating the genes required to make the *H. influenzae* (Hib) capsular polysaccharide. The excitement of this new genetics was brilliantly captured in a seminal article in 1988 by the legendary microbiologist, Stanley Falkow, who traced the history of how Koch's nineteenth-century idea — the isolation of pathogenic bacteria, growing them in culture and then reproducing infection though animal experiment — was reformulated a century later through molecular biology. Its impact was breathtaking; the genes required for bacterial virulence could be isolated, inactivated or modified and then reinserted into a bacterial cell. This allowed scientists to pinpoint the role of specific gene sequences in different infections, including meningitis.

For me, researching the genes for the type b capsule and their role in meningitis was truly thrilling. It was also instrumental in my appointment to Oxford where the medical school had set its sights on cutting-edge research that would make it a world leader in molecular medicine, one thread of which I spun off into a major commitment to vaccines. So it was that when Craig Venter and Ham Smith joined forces in 1995 to sequence all the nucleotides of the *H. influenzae* (Hi) genome, my Oxford laboratory became a part of this revolution in biology. It unlocked the entire genetic pathway of endotoxin biosynthesis in *Hi* and meningococci and provided new insights into the evolution of mechanisms through which pathogens adapt to their host and evade immune clearance mechanisms and treatments, including vaccines and antibiotics. Using the power of the whole genome sequence, a "yellow pages" inventory of every potential vaccine antigen, this new technology was used to develop a vaccine against the elusive MenB bacterium. Meantime, I had established a clinical trials research platform through the Oxford Vaccine

Group (OVG) and set up the purpose-built Centre for Clinical Vaccinology. OVG was responsible for more than half of the children enrolled in the trials of the 4CMenB vaccine that was eventually approved in 2013. It is not so often that one has the privilege of seeing a project go from concept to clinical implementation. Since 2014, all infants in the UK have been offered a vaccine that has had a substantial impact on preventing cases of meningococcal sepsis and meningitis.

So, where do we stand today in the battle against bacterial meningitis? There has been a halving of cases of Hi-b meningitis worldwide, but these vaccines have not yet been introduced in China, Russia or Thailand. My colleague Mathuram Santosham[b] spoke to me of the irony he felt when, as one of the recipients of the Prince Mahidol Award in 2019, he gave his award lecture from the podium of the Grand Bangkok Convention Centre in a ceremony that was attended by Thailand's Minister of Health!

Undoubtedly, one of the biggest success stories of immunisation in the twenty-first century is the implementation of vaccines against epidemic Meningococcal A (MenA) infections in sub-Saharan Africa's meningitis belt. After so many years of the tragic carnage of repeated waves of meningitis, the introduction of the conjugate vaccine in 2010 has reduced its impact to the point where there are grounds for supposing that these capsular variants of the meningococcus may have been almost eliminated. But the meningococcus is nothing if not resilient. MenA may be vanishingly rare, but other capsular variants have emerged. Where MenA once ruled the roost, MenW, MenC and MenX capsular variants now dominate in Africa — for which different conjugate vaccines are needed.

The impact of the conjugate vaccines on pneumococcal meningitis is even more of a mixed message of success. Although conjugate pneumococcal vaccines have been highly successful in preventing pneumonia and other serious diseases, the same cannot be said for meningitis. Pneumococcal meningitis continues to cause morbidity and mortality among children and adults despite widespread use of pneumococcal conjugate vaccines in several countries around the globe. The reason for this is a dramatic change in the serotypes of pneumococci causing meningitis. The vaccines contain only a subset of the

---

[b] See Chapter 11.

many pneumococcal capsular types and since their introduction, pneumo-coccal meningitis is now caused by bacterial variants that have taken over the niche created by the almost complete elimination from the upper airways of the capsular types contained in the vaccine. This replacement is a sobering example of a phenomenon that has been given much emphasis elsewhere. I mentioned it first as the Professor's question in the context of *Hi*-b vaccines and again as the "elephant in the room" during the development of the Meningococcus B vaccine. It's a striking example of Darwinian theory: vaccines exert selection pressures that alter microbial populations. Why this has had such profound effects on pneumococcal, but not *Hi*-b or meningococcal conjugate vaccines, is not understood.

Another as yet unmet challenge is neonatal bacterial meningitis — strictly defined as an infection occurring within the first month after birth. Unlike older infants, young children and adults, the main causes of neonatal meningitis are encapsulated streptococci and a particularly virulent form of *Escherichia coli.* The approach to their prevention necessarily must be radi-cally different. Neonatal meningitis strikes often within hours of birth, so the ideal strategy is maternal immunisation. If the vaccine is given to the mother before she gives birth, the new-born baby can be protected by this passive transfer of immunity via the placenta — an area of current research that is beyond the scope of this book.

To paraphrase Winston Churchill: "this is not the end, not even the begin-ning of the end; but it may be the end of the beginning." WHO has established a global road map, ambitiously called "Defeating Meningitis by 2030," whose first priority is to continue the unfinished task of eliminating bacterial men-ingitis across the globe. While global meningitis deaths decreased by 21.0% from 1990 to 2016, the overall burden of meningitis remains high. Progress in reducing mortality and morbidity from this group of infections has sub-stantially lagged behind that for other vaccine-preventable diseases such as measles, tetanus and diarrhoeal disease.[1]

Beginning with the microscope and the discovery of bacterial cells, the scientific understanding of what causes so many of our planet's most serious diseases has changed how we live and our understanding of all life forms and their interconnections. Only when we regard science as indispensable and commit to share and communicate its potential with all our fellow citizens can we relate to the universe around us and acquire the intellectual framework for the development of our modern civilisation. Despite the extraordinary achievements in medicine since the discovery of germ theory around 150 years ago, infectious diseases remain responsible for about half of all deaths in humans. Moreover, it must surely weigh heavily on our minds that deaths and disabilities from infections are very unevenly distributed, depending on where you live and how wealthy or poor you are.

It is a challenge that is especially relevant to clinician-scientists who bring unique perspectives to medical research. The fragility of health and how it can give way to illness provides an inspirational and compelling opportunity to understand human biology. It is through nature's "experiments" that clinicians must confront the profound questions posed by our patients at the bedside and in the clinic. Whatever kind of research we do, it is imprinted by having cared for the sick. It is why as a clinician-scientist I believe that going back and forth between clinical medicine and (in my case) laboratory science is natural and almost necessary. But many find this constant transitioning to be a *pons asinorum*, a bridge too far given that either clinical practice or medical research is in its own right enormously challenging. As technology has advanced the reach of both, the feasibility and future of the careers of clinician-scientists have been and remain a subject of debate and concern. If this book has achieved one of its aims, it will be clear that I remain a passionate advocate of combining medical practice and original scientific research. My own experience has been personally so immensely rewarding and meaningful.

Medicine is learned by the bedside … See, and then research … But see first.

Reference

[1] *Defeating meningitis 2030: baseline situation analysis.* https://www.who.int/publications/m/item/defeating-meningitis-2030-baseline-situation-analysis.

Glossary

Acute. Used in medicine to describe a disease or disorder that comes on rapidly, is accompanied by distinct symptoms and is usually of short duration.

Active surveillance. A form of epidemiological tool, often used in epidemics or pandemics, to enhance the reporting of the number of cases of disease and outcomes including deaths and disabilities. It requires substantially more resources than *passive surveillance* where there is no stimulus, such as reminders or feedback, to facilitate or incentivise the involved health authorities.

Adaptive immunity. The form of acquired immunity that combats germs through either antibodies (made by B lymphocytes) or by killing infected host cells (though T lymphocytes).

Altruism. Disinterested and selfless concern for the well-being of others.

Ampicillin. An antibiotic drug (chemically related to penicillin) used for treating a wide range of bacterial infections.

Anthrax. Rare but serious illness caused by the bacterium *Bacillus Anthracis*. It mainly affects farm animals, such as sheep, but humans can become infected through contact with these animals. Causes skin sores and a generalised illness with vomiting and collapse. The bacterium is also of concern as a potential weapon of bioterrorism.

Antibody. A Y-shaped protein of the immune system that neutralises or facilitates removal or killing of a pathogen. The tips of the "Y" are specific for a small region of an *antigen* resulting in a unique "lock and key" type of recognition.

Antigen. Any substance that induces the immune system to produce antibodies, most commonly a molecule composed of protein, sugar or lipid, on the surface of a germ.

Antigenic variation. The mechanism by which a microbe alters its surface molecules. The resulting variants may avoid natural or vaccine-induced immune responses.

Anti-vaxxer. A person who is opposed to vaccination, typically a parent who does not wish their child to be immunised.

Attenuation. Reducing the severity, virulence or infectiousness of a germ. For example, the modification required to make a bacterium or virus that is safe for use as a vaccine.

Autism. A developmental disorder characterised by difficulties with social interaction and communication.

B-cell (also known as B lymphocyte). A specific type of white blood cell that produce antibodies, a major part of the adaptive immune system.

Bacillus subtilis. A soil bacterium that has achieved notoriety among scientists because of the insights it has provided into formation of spores, a dormant state of bacteria with reduced metabolism and respiration.

Bacteraemia. The presence of bacteria in the blood often referred to as *blood poisoning.* See also *septicaemia.*

Bacteria. A microbe that thrives independently in diverse environments. It is a single cell surrounded by a membrane that encloses cytoplasm and genetic information in the form of DNA.

Bacterial Polysaccharide Immune Globulin (BPIG). A form of treatment consisting of antibodies prepared from the plasma of donors immunised with *Haemophilus influenzae* type b, pneumococcal and meningococcal vaccines used to prevent infection by these organisms in high-risk patients who have not or cannot be immunised.

Bactericidal activity. A general term for the killing of bacteria.

Bactericidal assay. A laboratory test used to measure the potency of immune factors (usually antibodies) in killing bacteria.

Bacteriophage. Also known as a *phage*, a bacteriophage is a virus that infects and multiplies within bacteria.

Biochemistry. The study of chemical processes within and relating to living organisms.

Brain Fog. A form of lingering brain malfunction experienced after infection with COVID-19 that includes loss of memory, confusion, inability to concentrate, alterations in emotional state and behaviour.

Carrier. A person who harbours a pathogen, often without experiencing signs or symptoms of infection and who can serve as a potential source of infection to others.

Carrier protein. When used in the context of conjugate vaccines, refers to the protein or peptide that is chemically linked to a sugar molecule (polysaccharide or oligosaccharide) to enhance the production of antibodies through cooperation with T-cells.

Centre for Clinical Vaccinology and Tropical Medicine (CCVTM). A purpose-built facility at the University of Oxford dedicated to vaccine-related and worldwide research on infections.

Cerebral cortex. The outer layer of the brain. It is covered by the meninges and is often popularly referred to as "grey matter."

Cerebrospinal fluid (commonly abbreviated to CSF). The clear fluid that surrounds the brain and spinal cord. It acts as a cushion to protect the brain from injury and provide nutrient delivery and waste removal from the nervous system.

Chloramphenicol. An antibiotic with a broad spectrum of activities for treating bacterial infections.

Choroid plexus. The part of the brain that produces cerebrospinal fluid (CSF). It comprises small blood vessels lined by specialised cells (*ependyma*). It serves as a cellular barrier between the blood and the *subarachnoid space* that contains the CSF to protect against harmful germs and toxins.

Chronic fatigue syndrome. A long-term illness of uncertain cause characterised by extreme fatigue lasting at least six months and often longer.

Clapier. French slang for "brothel" and possibly the etymological origin of "the clap," a familiar name for infection with the gonococcus, a bacterial sexually transmitted disease.

Clinician-scientist. A person with a degree to practice medicine who spends significant time and professional effort in scientific research and therefore correspondingly less time in clinical practice compared to other physicians.

Colonisation. The presence and growth of microbes on and within body sites that are acquired through exposure to the environment, including other persons, animals etc.

Community protection. See *herd immunity*.

Complement pathway. An integral part of the immune system that enhances the ability of antibodies to remove or kill pathogens or damaged cells.

Contagion. The spread of an infection by transmission of a pathogen from one person to another.

Cowpox. A bovine virus that is closely related to the vaccinia or smallpox virus. It can infect humans and stimulate immunity to smallpox.

Cytokine. A peptide that is released by cells that have signalling properties, for example, to the immune system, to coordinate biological responses against infection and trigger inflammation.

Cytoplasm. The semi-fluid liquid within a cell, enclosed by the cell membrane. It is made up largely of water, salts proteins, fats (lipids) and nucleic acids.

Delirium tremens. Often referred to as "the DTs." A severe form of alcohol withdrawal manifested by mental confusion and altered, sometimes violent, behaviour.

DNA. The abbreviation of **d**esoxyribo**n**ucleic **a**cid. The hereditary material in almost all life forms that is the chemical basis of genes.

DNA library. A collection of the total genomic DNA from a single organism. Each fragment of the collection of DNA fragments is inserted into a special virus so that, in aggregate, the library possesses at least one representative sequence of the entire genome.

DNA probe. A small single strand of DNA sequence used to detect the presence of an identical sequence by relying on the complementary pairing of nucleotides (Watson–Crick base-pairing), known as DNA hybridisation. Detection uses various methods such as radioactive labelling.

Drosophila melanogaster. A species of fly, often referred to as a "fruit fly." One of the commonest model organisms used in biological research, especially genetics.

Encephalitis. Inflammation of the brain caused by an infection (e.g., a virus) or through the person's own immune system attacking the brain.

Encephalitis lethargica. An atypical form of encephalitis that spread across the world coinciding with the influenza pandemic known as the Spanish flu.

Encephalopathy. An altered state of brain function induced by damage or disease.

Endotoxin. A component of the cell wall of certain bacteria. It is often called lipopolysaccharide (or lipo-oligosaccharide). It consists of an innermost Lipid A portion containing fatty acids and sugar (disaccharide) phosphates attached to the outermost core sugars. See Figure 11.1.

Enzyme. A protein that regulates the rate at which chemical reactions occur without being altered in the process.

Ependyma. A thin layer (membrane) of specialised cells which line spaces within the brain and spinal cord (spinal canal) that contain cerebrospinal fluid (CSF).

Epidemiology. The discipline for the study of the distribution (temporal and geographical) and determinants (causes and risk factors) of diseases in specified populations.

Escherichia coli. A diverse group of bacteria that are normal inhabitants of people and animals. Some types cause disease, but most are harmless and help keep the body healthy. Used as model organism in biological research, especially genetic and recombinant DNA experiments.

Expressed Sequence Tags (ESTs). Genetic constructs made by converting the *messenger ribonucleic acid (mRNA)* for making proteins into DNA. This pioneering discovery by Craig Venter and his team at the National Institutes of Health, USA, provided insights into the subset of genes that were actively expressed in particular human cells.

Febrile seizure. A convulsion or fit in a child (aged from about 3 months to 6 years) caused by a fever. Most are harmless and are followed by a full recovery.

Fermentation. The process of converting carbohydrates (complex sugars) to alcohol using microbes, including bacteria and yeasts, under conditions that lack oxygen.

Floppy disc. A magnetic disc of a thin and flexible material used for storing information on and requiring the use of computers.

Fungi. A group of organisms that include yeasts, moulds and mushrooms. Among the microscopic forms are various species that cause infections, such as *Candida albicans,* commonly known as "thrush." One species, *Cryptococcus neoformans*, is an important cause of meningitis.

GAVI. An acronym for the Global Vaccine Alliance, an international organisation created in 2000 to improve access to new and underused vaccines for children living in the world's poorest countries.

Gene. A region of DNA, varying in size from a few hundred to more than two million nucleotides, that contains information for making the proteins that affect an organism's function (often called a "trait" or "phenotype"). The DNA sequence of a gene or "genotype" is inherited by offspring and, through natural selection, is a fundamental unit of the evolution of all life forms.

Genetics. The branch of biology concerned with the study of genes.

Genome. An organism's complete set of genetic instructions for making different proteins.

Genotype. In a broad sense, refers to the genetic make-up of an organism, but also refers to the information contained within each of an organism's genes.

Glycoprotein. Proteins that also have an attached carbohydrate (made of multiple sugars) that form a hybrid molecule. For example, the chemical bonding of capsular polysaccharides to carrier proteins used in making "conjugate vaccines."

Gonorrhoea. Colloquially known as "the clap" is a sexually transmitted infection caused by the bacterium *Neisseria gonorrhoea*.

Hapten. A small molecule that elicits an immune response only when attached to a larger molecule (see *carrier protein*).

Herd Immunity (also *community protection*). Resistance to the spread of a specific infectious disease (e.g., measles, whooping cough) that occurs when a sufficient proportion of individuals in a population are immune through previous infection or immunisation.

Hesitancy (vaccine). Refers to refusal or delay in acceptance of vaccines despite availability of immunisation services.

Hookworm. A common intestinal parasitic worm found in humans. The eggs of the parasite enter the body from the soil. Heavy infection causes abdominal pain, diarrhoea, loss of appetite, weight loss. It is one cause of severe anaemia (reduction in red blood cells).

Il Palio. A horse race that takes place in the central plaza in Siena, dedicated to the Virgin Mary, that is held twice a year, in July and August. Ten horses and riders, bareback and dressed in appropriate colours, represent ten of the 17 city neighbourhoods (*contrade*).

Immunity. The ability of an organism to resist infection or injurious substances released by pathogens through a variety of intrinsic or acquired host mechanisms.

Immunodeficiency. An impaired state, partial or complete, in which the immune systems of the body are unable to fight infections and other diseases, such as cancer. It can be present from birth or acquired later.

Immunological memory. The ability of the immune system to respond more rapidly and effectively to pathogens that have been encountered previously.

Immunology. The branch of medicine and biology concerned with immunity.

In silico. Experimental studies in the sciences that are conducted using a computer.

In vitro. Experimental studies in the sciences that are conducted using components of an organism that have been isolated outside of their normal biological context.

In vivo. A process or experiment taking place in a living organism.

Infant. Strictly defined as a baby in the first year of life, but mostly used as a general term for a young baby.

Inflammation. The general term for the complex biological response of body tissues to harmful stimuli including pathogens, damaged cells or irritants.

Innate immunity. Immune responses that can be activated immediately against infection that are not specific to any one pathogen (see also and contrast with *adaptive immunity*).

Institut Pasteur. French (Parisian) biomedical research foundation known for its extraordinary contributions to microbiology and infectious diseases that were made after Louis Pasteur's death after whom the Institute was named. Its research has included seminal research on diphtheria and antitoxins, plague, BCG vaccine for TB, phagocytes, polio virus, bacteriophages, antibodies and complement, typhus, yellow fever vaccine, sulphonamides, gene regulation and Human Immunodeficiency Virus (HIV). Pasteurian scientists have won ten Nobel Prizes.

International Pathogenic Neisseria Conference (INPC). A biannual scientific meeting that is dedicated to research on the gonococcus and meningococcus. The first meeting was held in 1978 in San Francisco, followed since by meetings in 20 other different venues in the United States, Canada, Australia, and Europe.

Intern. A medical student or trainee who works to gain experience and/or to satisfy requirements for a qualification.

Isogenic. Organism having the same or nearly identical genotypes.

Isomer. Compounds that have the same chemical formula but spatially distinct arrangements of atoms.

Lasker Awards. Named after Albert Lasker (1880–1952), an American businessman and philanthropist, whose foundation is responsible for prestigious awards for discoveries that open up a new area of biomedical science.

Licensure (of a vaccine). The complex process required by regulatory authorities to ensure that a vaccine meets strict standards of safety, quality and effectiveness.

Long COVID. The term that describes the effects of infection with COVID-19 virus that continue for weeks or months beyond the initial illness.

Lumbar puncture. A medical procedure (also known as a "spinal tap") in which a needle is inserted into the spinal canal to obtain cerebrospinal fluid for diagnostic testing. It is one of the most important methods of diagnosing meningitis and other kinds of brain inflammation.

Lymph node. Lymph nodes are small round, bean shaped glands distributed throughout the body. They consist of different types of cells that filter out foreign substances and initiate immune responses.

Lymphocyte. Type of white blood cell that is of fundamental importance to the immune system. There are many different types with different functions. For example, B lymphocytes make antibodies; T lymphocytes potentiate antibody production; etc.

Lysis. The disintegration of a cell by rupture of the cell wall or membrane.

Macrophage. Specialised cells involved in the detection, phagocytosis and destruction of bacteria and other pathogens.

Meningitis. Inflammation of the linings of the brain.

Meningitis Now. A national UK charity formed following the merger of the Meningitis Trust and Meningitis UK in 2013. Their vision is a future where no one in the UK loses their life to meningitis and everyone gets the support they need to rebuild their lives.

Meningitis Research Foundation. A UK-based charity dedicated to research, education and outreach support relating specifically to bacterial meningitis. Responsible for the action plan, endorsed by the World Health Organisation, "Defeating Meningitis 2030."

Meningococcal Antigen Typing Scheme. An assay based on genomics used to predict the potential protection of the meningococcal (4CMenB) vaccine against individual isolates of meningococci isolated anywhere in the world.

Methotrexate. A drug used to treat cancer and rheumatoid arthritis. It works by interfering with folic acid, an essential factor required for the maintenance and replication of cells.

Microbiome. The genetic material of the microbes (bacteria, fungi, viruses, protozoa) that live on and inside the human body.

Microsatellites. Sequences of non-coding repetitive DNA used as genetic markers to follow the inheritance of genes in families.

Molecular weight. The average weight (mass) of a given molecule calculated by adding together the masses of its constituent atoms.

Monoclonal antibody. An antibody derived from a single B lymphocyte that recognises a specific region of an antigen.

Mycobacterium tuberculosis. The bacterium of ancient origin often called consumption or simply TB. It is one of the world's most important causes of pneumonia and meningitis and was responsible for killing around 1.5 million in 2020.

National Childhood Encephalopathy Study (NCES). A case-controlled study of neurological illnesses in children conducted in 1981. Its main aim was to investigate the relation between immunisation with pertussis vaccine (administered along with diphtheria and tetanus vaccines) and brain damage. The findings of the study were inconclusive.

National Institute for Clinical Excellence (NICE). An executive, non-departmental body of the Department of Health (England) that publishes guidance on health technologies, clinical practice, health promotion and social care.

Neutrophils. White blood cells that are capable of ingesting and killing bacteria (and other microbes). They are one of the earliest cell types to travel to sites of infection in the body, a first line of immune defence against infection.

New genetics. A term used to highlight how recombinant DNA technology, through identifying variations in the DNA sequence of genes, changed the opportunities for a better understanding of disease.

Nuclear Magnetic Resonance (NMR). A technology in which nuclei are studied in a strong magnetic field to investigate the structure of proteins and other complex molecules.

Oedema. An abnormal build-up of fluid in the body that causes swelling.

Olfactory. Refers to the sensory nervous system for smelling.

Oxford Vaccine Group. A vaccine research group within the University of Oxford Department of Paediatrics. Founded in 1994 and now located in the Centre for Clinical Vaccinology. Its founding Director was Richard Moxon and its current director is Andrew Pollard.

Parasite. A general term that describes organisms that live on or in an organism, but often used to denote specifically various protozoa (includes malaria), worms and ectoparasites such as fleas and mites.

Passive immunisation. Protection against disease that is brought about by transferring antibodies from another person rather than by the person's own natural or induced antibodies.

Passive surveillance. See *active surveillance*.

Pathogen. A germ (microbe) that causes disease. The general term includes many species of viruses, bacteria, fungi and protozoa.

Pathogenicity. Synonymous with virulence: The ability and degree to which a microbe causes damage to an infected host.

Pébrine. So-called "pepper disease" of silkworms caused by a parasite, most commonly *Nosema bombycis*.

Pertussis. Commonly called whooping cough, a highly contagious respiratory disease caused by the bacterium *Bordetella pertussis*.

Petri plate (or dish). A shallow transparent lidded dish that biologists use for growing cells, such as bacteria, on a semi-solidified growth medium.

Phenotype. An organism's observable traits, such as height or eye colour, that result from the interaction of genotype with environmental factors.

Photosynthesis. A process used by plants and other organisms to convert light energy into chemical energy requiring water and carbon dioxide.

Phrenitis. An old-fashioned Greek term (origin of our word "frenzy") for inflammation of the brain or other parts of the body.

Phylogenetic tree. A branching diagram to show the evolutionary relationships within and between various species of organisms.

Placebo. A substance which is designed to have no treatment value. In trials, it can be made to resemble an active treatment so that it functions as a "control."

Plasma. The straw-coloured liquid component of blood that remains when blood cells are removed. It contains water, salts, enzymes and various proteins such as antibodies.

Plasmid. A small DNA molecule that is physically separated from the chromosome. It can replicate itself independently.

Polymer. A substance composed of many identical or non-identical molecules that are linked together.

Polymorphonuclear leucocyte. A white blood cell originating from the bone marrow with a multi-lobed nucleus that plays an important part in the innate immune response to microbes or other noxious agents. Also often called a granulocyte.

Polyribose-ribosyl phosphate. The chemical basis of the polymer of the type b capsular (polysaccharide) antigen of the bacterium *Haemophilus influenzae* (*Hi*-b).

Polysaccharide. A large molecule made of many smaller sugar units that are joined together in a long chain.

Precipitin reaction. The outcome of the interaction between an antibody and an antigen leading to an insoluble complex.

Protective efficacy. The percentage reduction of a disease resulting from an intervention, for example, a vaccine.

Protein. A complex substance that consists of amino acids chemically linked together in a folded chain of three dimensions. Proteins are found in all living organisms and are an essential basis of their biological functions.

Protozoa. A single cell organism with a membrane-bound nucleus, for example, amoebae or malaria parasites.

Puerperal sepsis. The infection of the female genital tract occurring at labour or within 42 days after giving birth.

Recognition. A technical biological term used to describe the identification of molecules that are non-self during antigen presentation.

Recombinant DNA. The technology of joining together DNA molecules from two different species. The resulting hybrid molecules can be inserted into an organism, such as a bacterium, to produce novel attributes of value to medicine, agriculture and industry, etc.

Restriction enzymes (endonucleases). A DNA cutting enzyme that recognises a specific region (sequence of nucleotides). They are found in bacteria where they act as a defence mechanism against invading viruses. Their use in mapping, cloning and many other DNA technologies has revolutionised molecular biology.

RNA. The abbreviation of Ribonucleic acid. A nucleic acid polymer, a molecule essential in various biological roles in coding, decoding, regulation and expression of genes. It differs from DNA in its constituent sugar, ribose (DNA contains deoxyribose that lacks one oxygen atom), and in using the nucleotide uracil as compared to the thymine in DNA.

Russian flu. Refers to the influenza pandemic of 1889–1890.

Sepsis. A life-threatening illness and response to infection that can lead to tissue damage, organ failure and death. When associated with bacteraemia, the condition is called *septicaemia.*

Septicaemia. A serious, life-threatening illness in which there are bacteria in the blood stream. Also often called blood poisoning.

Serum. The clear yellowish fluid that remains after clotting factors and cells have been removed from blood. The removal of blood cells only is called *plasma*.

Single cell bottleneck. The phenomenon in which a population of individual organisms is reduced to just one. It can occur through selective or non-selective mechanisms and has important implications for the genetic diversity of populations of life forms.

Smallpox. A highly contagious and deadly virus that affected humans for thousands of years. There was no known cure until the development of vaccines that eliminated the virus from the planet (1977).

Spanish flu. Refers to the influenza pandemic that was at its height in the years 1918–1919.

Species. A biological classification that identifies related organisms into groups that share common characteristics. The term is complex and controversial because of the numerous methods (ranging from morphology to DNA sequence) that are used to define the basis of a *species*.

Specific soluble substance. The term used to refer to the capsular polysaccharide isolated from the pneumococcus and found in human body fluids (e.g., urine and blood) of patients with pneumococcal disease (e.g., pneumonia and meningitis).

Spike protein. A large surface exposed protein of the COVID-19 pandemic virus that is a major factor in the attachment (docking) of the virus to human cells through a receptor called ACE2. This protein is the key antigen of most of the current vaccines against COVID-19.

Spinal tap. Synonymous with lumbar puncture.

Subarachnoid space. The narrow space located between the two innermost of the three layers of the meninges that surround the brain and spinal cord. It contains cerebrospinal fluid (CSF).

Surveillance. In public health, the term used to describe the continuous, systematic collection, analysis, interpretation and analysis of health-related data.

T-cell (also known as a T lymphocyte). A white blood cell that plays a central role in the adaptive immune response. There are different kinds of T-cells, for example, those that help B-cells to make antibodies and those that can kill host cells that are infected with viruses.

Tight junctions. A structure, made mostly of proteins, that forms a barrier to block the passage of proteins and liquids traversing the space between adjacent cells.

Transfection. A method by which nucleic acids are translocated across the membrane of a cell to enter its cytoplasm and nucleus.

Transformation. A mechanism of transfer in which cell-free DNA is taken up and incorporated into the genome of a living cell, usually a bacterium, yeast or plant. When the recipient cell is a mammalian cell, it is often called *transfection*.

Translational research. The process of applying knowledge from basic biology to address critical needs in medical practice.

Vaccine. A substance used to induce immunity against disease, typically an infection, without causing the disease. It is prepared from the causative agent (usually modified) of the disease, one or more of its components or synthetic substitutes.

Vaccine Confidence Project. An initiative dedicated to monitoring public confidence in immunisation programmes worldwide. Its Director is Heidi Larson based at the London School of Hygiene and Tropical Medicine.

Vaccine failure. Describes an instance when a person contracts a disease despite being vaccinated against it.

Vaccine hesitancy. A reluctance or refusal to be vaccinated or to have one's children vaccinated.

Vaccinology. The science of vaccines, including their components (antigens), immune responses, delivery strategies, associate technologies, manufacturing, implementation, clinical evaluation and impact.

Variolation. The method of inoculation first used to immunise against smallpox using scabs or pustules from an infected individual that were scratched into the skin or inhaled into the nose.

Virulence. The ability and degree to which a microbe causes damage to an infected host. This is usually considered to be synonymous with *pathogenicity*.

Virulence factor. A pathogen-associated molecule that is required for it to cause disease.

Yellow fever. A serious virus infection, spread by mosquitoes, causing high fever, chills, headache muscle pains and sometimes jaundice (hence its name).

Printed in the United States
by Baker & Taylor Publisher Services